HOWARD W. SAMS OSCILLOSCOPE GUIDE

BY DR. ARNOLD J. BANKS

PROMPT© Publications is an imprint of Howard W. Sams & Company, A Bell Atlantic Company, 2647 Waterfront Parkway, E. Dr., Indianapolis, IN 46214-2041.

International Standard Book Number: 0-7906-1124-4
Library of Congress Card Catalog Number: 97-68180

Acquisitions Editor: Candace M. Hall
Editor: Loretta L. Leisure
Assistant Editors: Pat Brady, Natalie Harris
Typesetting: Loretta L. Leisure
Layout Design: Loretta L. Leisure
Cover Design: Christy Pierce
Graphics Conversion: Terry Varvel
Illustrations and Other Materials: Courtesy of the Author and the Tektronix Company

PRINTED IN THE UNITED STATES OF AMERICA

9 8 7 6 5 4 3 2 1

This book is dedicated to Birdie,

who would have given her all to live just

a little longer to help with the completion of this book.

I would also like to thank the following

family and friends for keeping me going when

I felt like giving up:

my mother, Ms. O. Banks,

my sons, Omar and Sharif,

my daughters, Lamar, Princess and Abreea,

and my life long friend, Marilyn.

And finally, thanks to the staff at PROMPT®

Publications and the Tektronix Company.

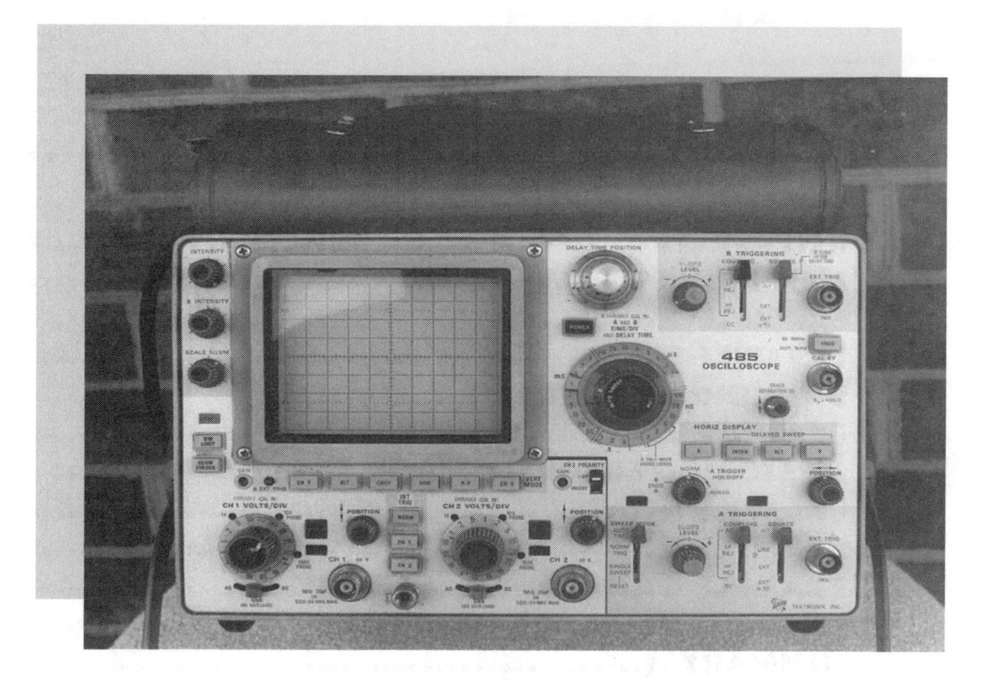

TABLE OF CONTENTS

CHAPTER 2

CHAPTER 3

CHAPTER 7

CHAPTER 8

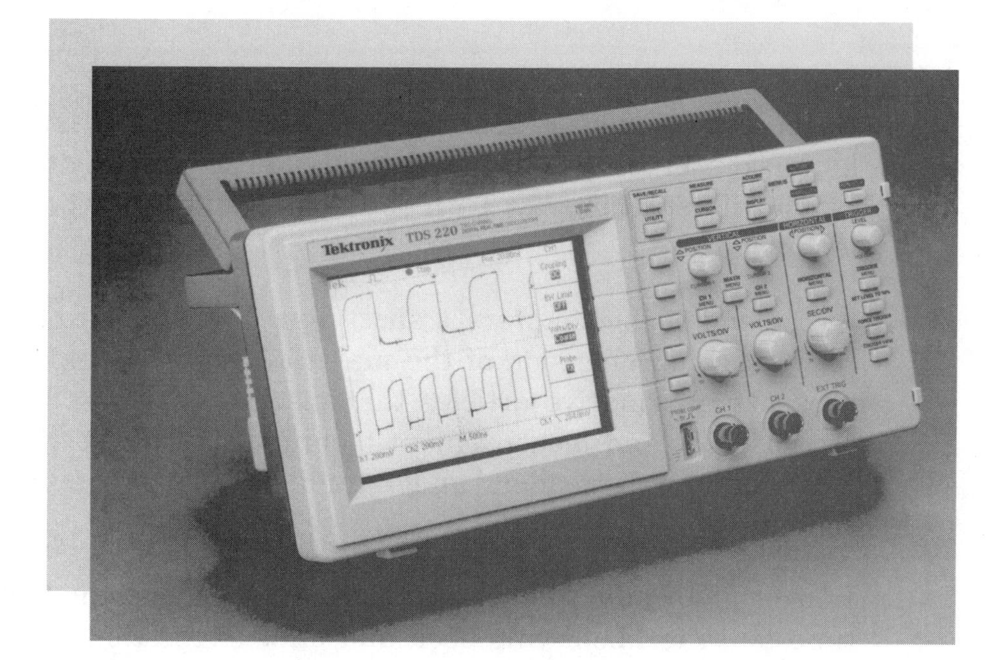

PREFACE

The purpose of *Howard W. Sams Oscilloscope Guide* is to provide a better understanding of using your oscilloscope to test and service electronic equipment. This book includes many hours of research, information and articles of importance to those who are planning on purchasing or setting up a test or service bench for repairs or general oscilloscope use.

Included in this book are hot tips on how to use your oscilloscope in making differential measurements, probe calibrations and voltage and frequency measurements using the oscilloscope's horizontal time base and voltage/div ranges.

The *Oscilloscope Guide* consists of nine chapters, a glossary of oscilloscope terms and extra articles and reference texts. This handbook serves as a complete guide of useful safety tips and helpful illustrations for both the novice and the experienced technician.

At the end of the chapters, a short quiz is provided to test the readers' comprehension of the material covered. Appendix A presents informative articles that discuss Tektronix's digital oscilloscopes and measurement techniques. This section will help heighten the readers' understanding of deeper oscilloscope concepts.

In closing, I would like to thank the many people who took the time to write about this book and offer suggestions for topic chapters. They also provided valuable and perceptive feedback on the material. They include: Denise Michalson, Edward Baca, P.S. Neelakanta, Raymond Bamberry of JPL, Ronald Fant of the Kessler Institute, Marilyn Pashby and Marjorie Yap of Tektronix Inc., and Charlie Lee of Sersa GeoComm.

We also wish to thank Candace Drake Hall, the managing editor at PROMPT, for her continuing support, encouragement and enthusiasm.

Arnold J. Banks, Ph. D.

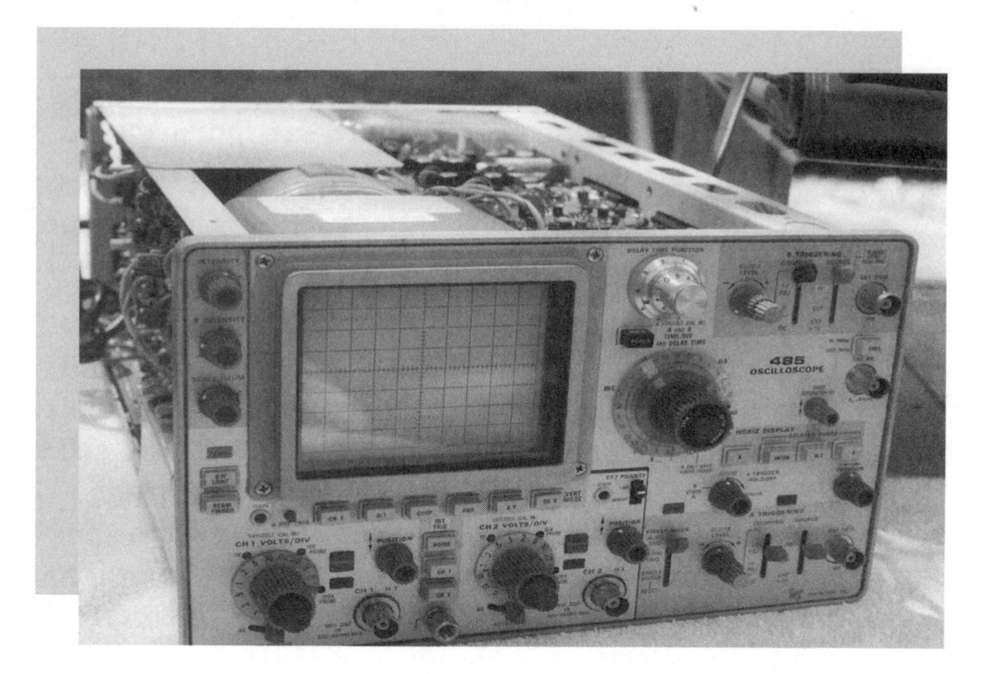

INTRODUCTION

If asked what would be the most essential tool for testing electronic equipment, electronic components, designing new technologies or just viewing the motion of thousands of particles vibrating in a maze of totally invisible patterns of elements in real time, the word oscilloscope would come to mind, and the adjectives that would describe this instrument would accurately characterize this marvel of engineering genius.

The word oscilloscope comes from the Latin root meaning *swing*, although the meaning of the first few letters of the word oscilloscope (i.e., oscillo) means to oscillate back and forth during a period of time, like a pendulum. The word *scope* is from the English root meaning an instrument for viewing.

Today the bench oscilloscope, portable oscilloscope, digital oscilloscope and scope meters prove to be an indispensable test tool for the scientist, engineer, electronic technician and student.

Today, small oscilloscopes weigh less than 5 pounds. The oscilloscope has found its use in all fields of science, physics, communications, and TV and radio services worldwide. The first pictures seen from the planet Mars on television would not have been possible if engineers and technicians didn't use oscilloscopes to design and test the Jet Propulsion Labs's *Pathfinder*. Without oscilloscopes the music of Mozart or Brahms could not be reproduced accurately. The music could not be put on CD's, or any other sound medium, in any sound studio without utilizing an oscilloscope of some type to display the recording equipment characteristics.

It was a group of French physicists who first used a similar instrument called an Oscillograph and so named it in the early turn of the century. While the oscillograph continuously gained popularity, especially in the fields of biology and physics, its uses in electronics to measure high-frequency waveforms were limited and virtually impossible to study in real time.

In the years to follow more efficient tubes were designed. This generated a need for a more powerful instrument that could both display high-frequency waveforms, and measure signals in real time of frequency and amplitude domains. While it is not known which American company produced the first commercial oscilloscope, (Dumont and Tektronix run a close 1st in this race) I'll put my money on RCA.

The first oscilloscopes that were commercially available were of the vacuum tube variety and at times could be mistaken for a small television set. The oscilloscopes of that time had limited frequency capabilities and weighed in excess of 90 pounds without any accessories. These large oscilloscopes dominated the markets for many years and they became commonplace in research labs, engineering departments and service facilities throughout the world.

In one such laboratory, Albert Einstein was employed under Robert Oppenheimer in the desert of New Mexico on a secret mission code named, "Manhattan Project". A scientist under Einstein used more than 150 oscilloscopes to study elementary particles and nuclear chain reactions. While engineers and technicians who were employed by the government in New Mexico designed electronic equipment using the oscilloscopes of that day, today this laboratory is still active with engineers and technicians using modern oscilloscopes, but don't go there trying to get pictures like I did; they're not interested.

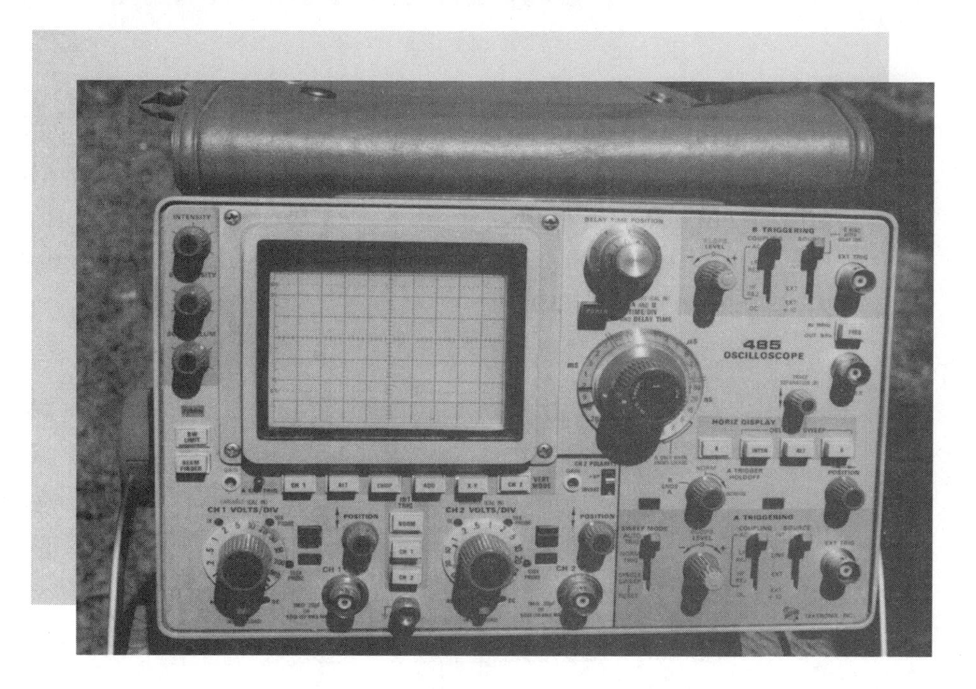

CHAPTER 1
OSCILLOSCOPE BASICS

CONTENTS:

The typical oscilloscope is an electronic test tool that displays electrical activity related to the electrical characteristics of the test application under concern. The electrical activity displayed on the cathode ray tube, (CRT) is called a *waveform*. It appears as a moving or stationary picture. **Figure 1-1** shows an example of a waveform.

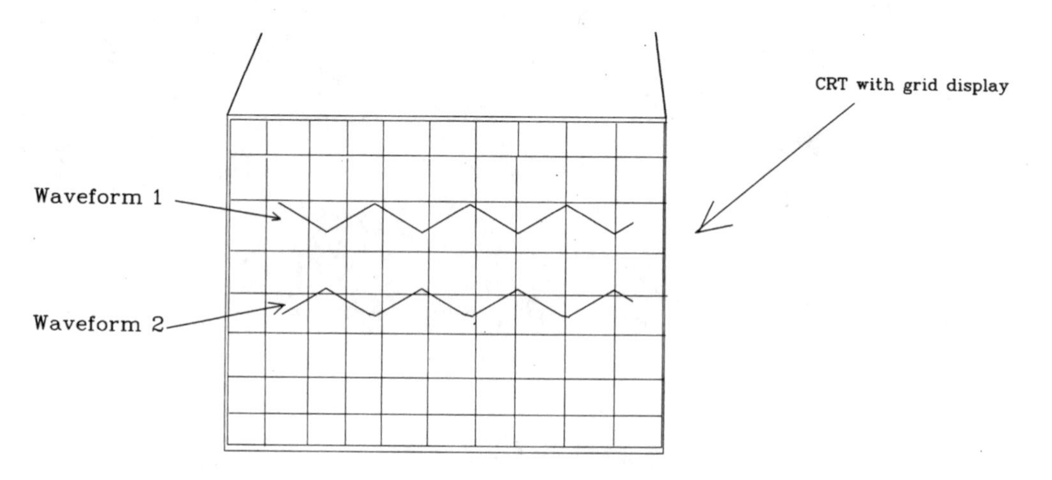

Figure 1-1. Simplified diagram of dual trace waveforms.

Except in a few cases of measurement, most waveforms are composed of three electrical components called *vectors*. You may remember vectors from math class. Each vector's direction can be drawn on the face of the CRT. Every waveform is composed of the components and direction shown in **Figure 1-2**.

The oscilloscope displays the information on the CRT when connected to the test circuit via a probe which transfers the electrical signal from the test circuit to the internal electronics of the oscilloscope. **Figure 1-3** shows a typical oscilloscope test application using a test probe to check the AC ripple across a 12 volt DC supply.

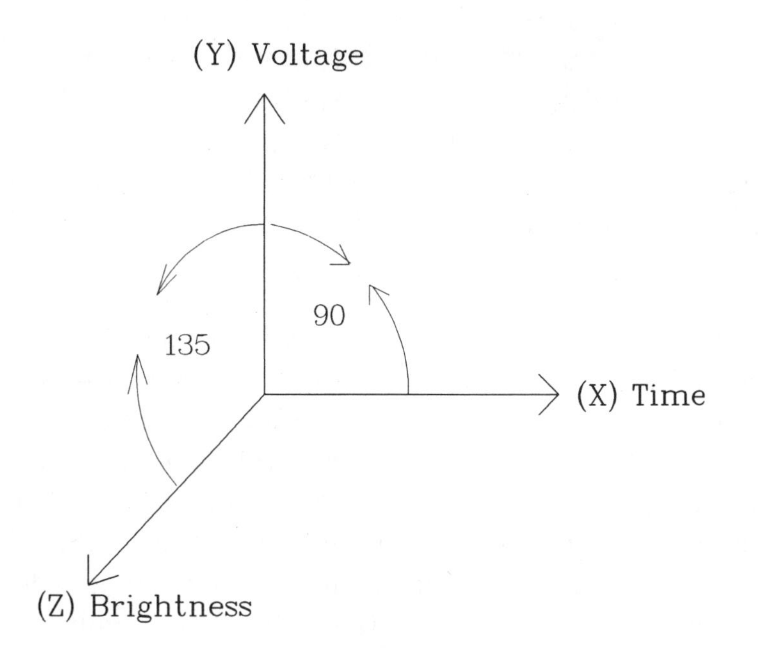

Figure 1-2. Vector illustration.

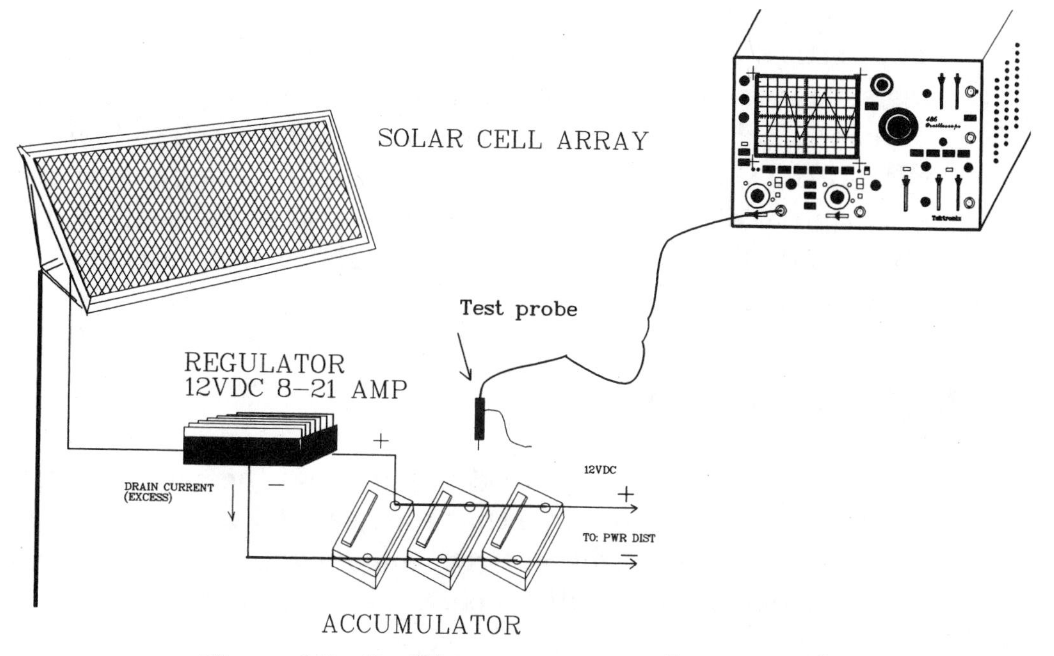

Figure 1-3. Oscilloscope connected to test probe.

There are five main areas that the user of the oscilloscope should become familiar with. These are the vertical section, horizontal section, triggering section, the CRT and the power supply. The CRT will be covered more thoroughly in the chapters to come. The power supply section will not be discussed in detail due to its large scope.

When the oscilloscope is connected to the test signal the voltage travels through the probe leads to the oscilloscope's vertical amplifier. There it is either amplified or attenuated by the volt/div control on the oscilloscope control panel; thus adjusting the voltage gain for the proper signal voltage level. The signal is then sent to the vertical deflection plates in the CRT. These plates cause the electron beam to move up and down on the CRT in relation to the signal amplitude and voltage polarity.

VERTICAL SECTION

The vertical section of the oscilloscope is responsible for making sure the test signals being analyzed or displayed are of the proper signal amplitude and frequency to be displayed on the CRT. If the vertical amplifier is saturated by the input signal the original electrical contents of the test signal will be distorted and the waveform will not be displayed accurately on the oscilloscope's display.

HORIZONTAL SECTION

It is the job of the horizontal section to move the beam in a left to right direction. The time that the beam travels from left to right is called the horizontal sweep and is controlled by the time base function on the oscilloscope.

When combined, input signals to the vertical and horizontal sections generate a waveform, as seen in **Figure 1-4**.

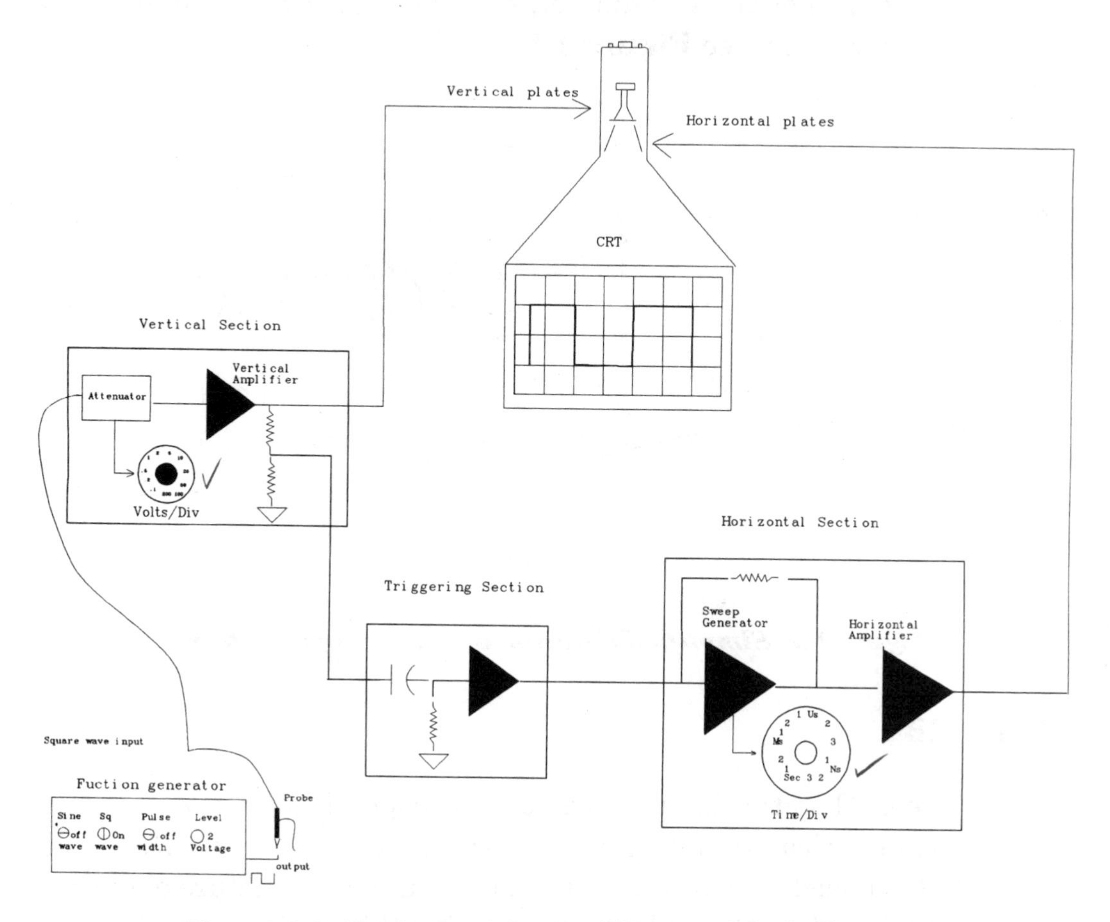

Figure 1-4. Typical analog oscilloscope block diagram.

TRIGGERING SECTION

At the beginning of this chapter we learned that a waveform consists of three components (X, Y, and Z) in order to properly view these components on an oscilloscope display we need to freeze the waveform.

The oscilloscope's triggering section allows the display waveform to become stationary by triggering the horizontal sweep at the proper time. This helps in the viewing and analysis of the waveform, see **Figure 1-5**.

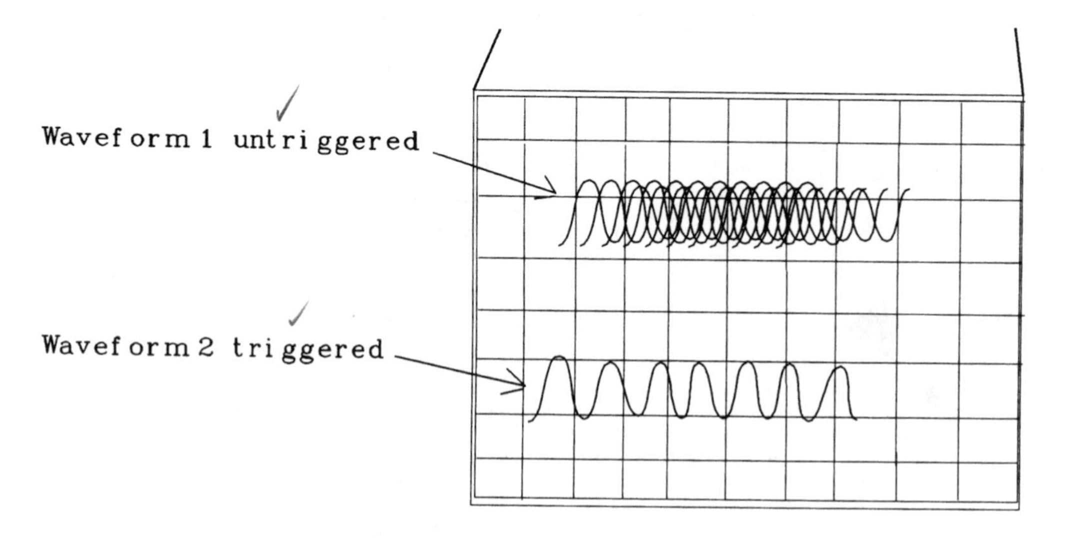

Figure 1-5. Simplified diagram of triggering waveform.

CRT TUBE

The CRT is the heart of the oscilloscope. It's job is to convert the electrical signal into a visual display (i.e., a picture). The CRT consists of a number of internal support structure elements and four deflection plates mounted within a glass envelope.

CRT GENERATED WAVEFORM

A fine beam of electrons is generated by emissions of a hot cathode that emits millions of electrons. These electrons bombard or impinge a cathodoluminescent screen surface in a glass envelope under a vacuum.

As each electron strikes the thin phosphor coating a small dot or beam of light is generated. The vertical and horizontal sections in the oscilloscope cause the signal voltage to deflect the electron beam in a fashion which follows the electrical characteristics of the circuit under test. See **Figure 1-6** for the CRT diagram.

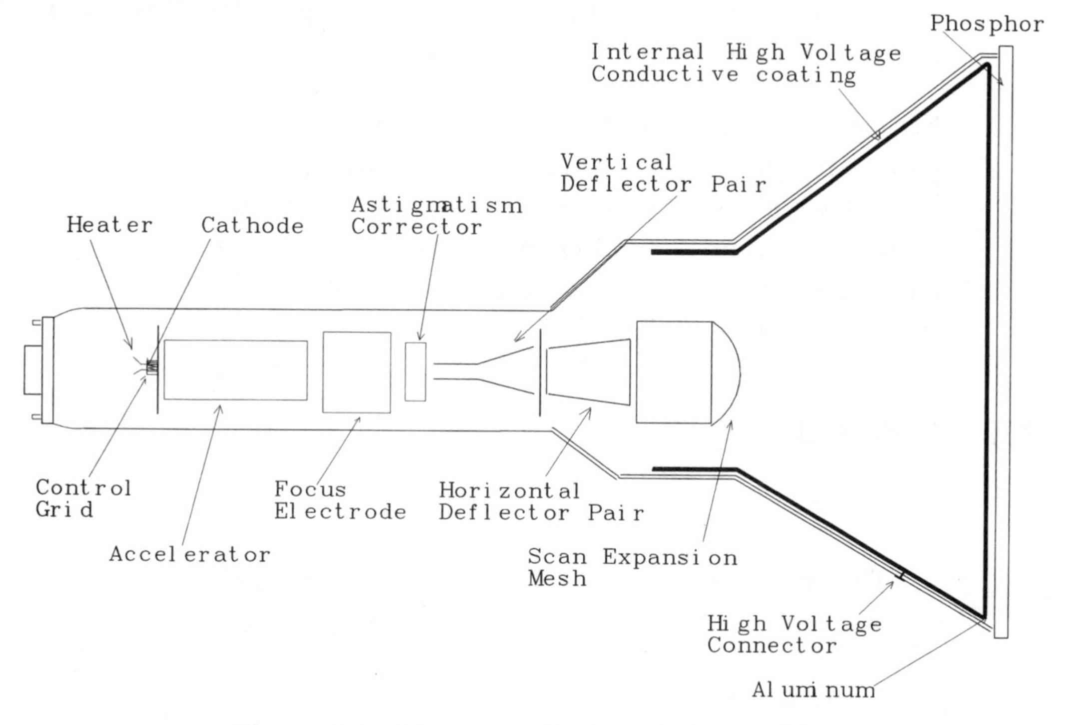

Figure 1-6. Diagram of internal view of CRT.

SUMMARY

The oscilloscope is a tool that generates a waveform when connected to a circuit for observation of that circuit's electrical characteristics. The five main sections of an oscilloscope are the vertical, horizontal, triggering, CRT and power supply.

For the oscilloscope to function as a test tool you must supply an input signal to the vertical amplifier via a probe.

The input signal must be of the proper amplitude and frequency in order to be displayed graphically on the oscilloscope display.

The trigger should always be adjusted to give the best signal representation on the oscilloscope display.

For additional information on basic oscilloscopes and/ or to help you better understand electrons, and how they interact in electronics we suggest you read *Basic Electronics* by Master Publishing, Inc., and *XYZ of Oscilloscopes* from Tektronix available free over the Internet @ http:// www.tek.com/Measurement/App_Notes/XYZs/contents.html.

CHAPTER 1 QUIZ

1. The interfacing device that connects the oscilloscope to the test circuit is a

   A. probe.
   B. test cable.
   C. test needle.
   D. none of the above.

2. How many components are in a waveform?

   A. 6
   B. 3
   C. 4
   D. 10

3. The direction of horizontal sweep on an oscilloscope is

   A. left to right. B. up and down.
   C. right to left. D. down and up.

4. The function of the vertical amplifier is to process

   A. the input signal. B. the frequency limit.
   C. the output signal. D. all of the above.

5. The main section of an oscilloscope a user does not come in contact with is the

   A. power supply. B. vertical section.
   C. triggering section. D. horizontal section.

6. The part of the CRT that emits electrons is the

   A. heat plate. B. grid.
   C. cathode. D. test probe.

7. The type of CRT coating which causes a light to be shown on the face plate when bombard with electrons is the

   A. cathodoluminescent. B. CAM.
   C. monochrome. D. epoxy.

8. ________ in a CRT causes the electron beam to move up and down.

   A. Potential uniformity B. Vertical deflection.
   C. Horizontal plates D. None of the above.

9. Which is the wrong signal characteristic for proper display viewing ?

   A. Frequency B. Signal level
   C. AC or DC D. None of the above

10. If an AC signal causes the beam to change its amplitude and frequency, how does a DC signal compare on an oscilloscope?

    A. Change of frequency
    B. High or low solid line
    C. Flickering trace
    D. Change of amplitude

11. What does the triggering control do?

    A. Cause the waveform to flicker
    B. Freeze the waveform
    C. Lock the CRT
    D. None of the above

12. What is the function of an oscilloscope?

    A. To create a moving picture in real time
    B. To display electrical activity ✓
    C. To record data
    D. None of the above

13. What is the function of the CRT?

    A. To convert electrical signals to visual information ✓
    B. To back-light the oscilloscope
    C. To make dots
    D. None of the above

14. What material is a CRT made of?

    A. Plastic
    B. Metal
    C. Glass ✓
    D. Wood

15. The signal source that must be connected to an oscilloscope input in order for oscilloscope to operate correctly is ______.

    A. AC
    B. DC
    C. AC or DC ✓
    D. Neither

NOTES

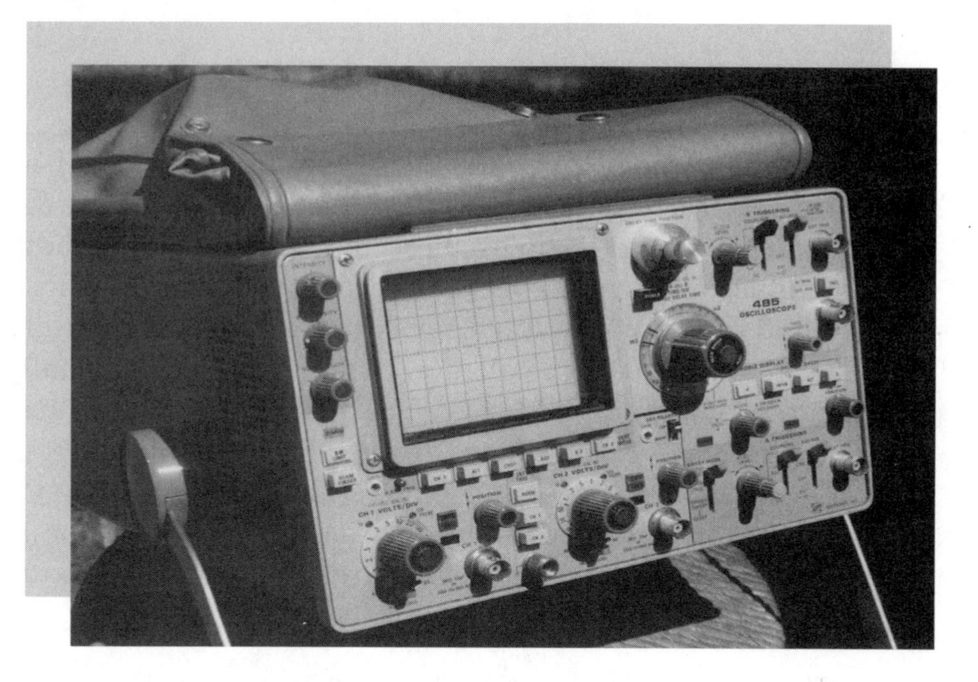

CHAPTER 2
OSCILLOSCOPE WAVEFORMS

CONTENTS:

INTERPRETING THE OSCILLOSCOPE WAVEFORM

In interpreting the oscilloscope waveforms the amplitude and time characteristics of the waveform are graphically displayed in square units. The oscilloscope's grid display familiarly resembles a chess or checker board that consist of a number of small square boxes. Each box is called a division. The grid lines form a grid pattern called a *graticule, which* is usually backlighted by the oscilloscope. The CRT graticule has an aspect ratio of 1:1.2. Each square graticule division measures 1 sq. cm per division and are used for reference points only on the oscilloscope display. Each square division is called a *major division.* Note: The oscilloscope viewing area measures approximately 8 cm vertically x 10 cm horizontally (2.7 in. x 3.2 in.)

The square boxes on the display form the grid which becomes transparent when power is applied. The shape and size of the grid, **Figure 2-1**, display a matrix that measures 8x10 (8 divisions vertically and 10 divisions horizontally).

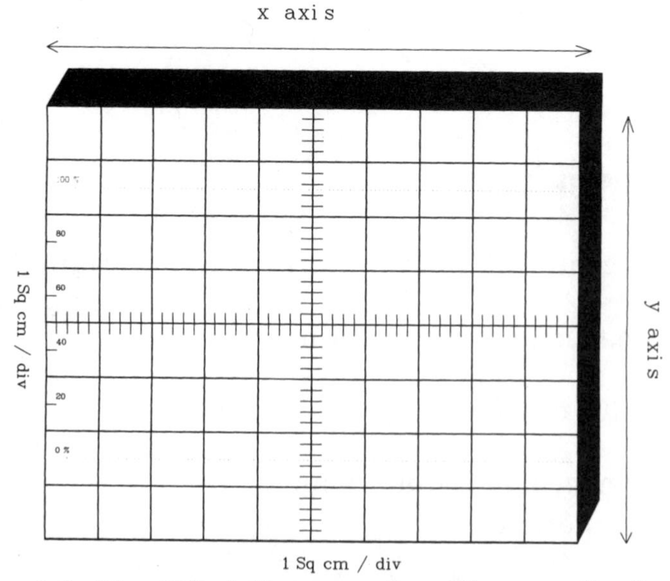

Figure 2-1. Simplified diagram of oscilloscope display grid.

MEASURING THE WAVEFORM AMPLITUDE

The waveform amplitude is measured by first counting the number of vertical divisions occupied by the waveform then multiplying the total number of vertical divisions by the Volt/Div control on the oscilloscope panel. See **Figure 2-2**.

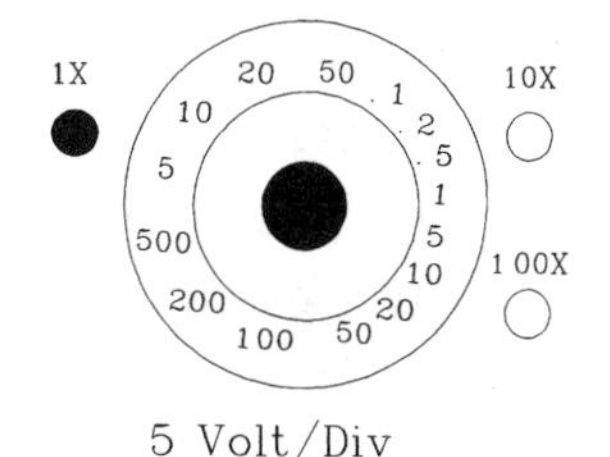

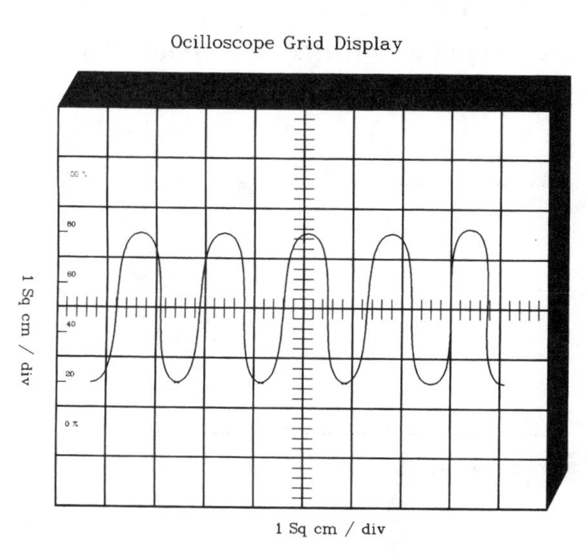

Figure 2-2. Display of waveform measuring 3 units vertically and a Vert/Div setting of 5 volts per division..

EXAMPLE 1:

The waveform in **Figure 2-2** measures 3 units vertically and the oscilloscope Volt/Div reads 5 volts per division. The signal voltage would be:

3 vertical units x 5 volts per division = 15 volts

EXAMPLE 2:

The waveform in **Figure 2-3** measures 5 units vertically and the oscilloscope Volt/Div reads 10 volts per division. The signal voltage would be:

5 vertical units x 10 volts per division = 50 volts

In both examples we call this voltage the *peak to peak voltage.* The peak to peak voltage measures the maximum and minimum peaks of a waveform. The *peak voltage* measures only the maximum peak or minimum peak of a waveform. Engineers and technicians use peak to peak voltages and peak voltages to describe oscilloscope measurements and testing of electronic equipment.

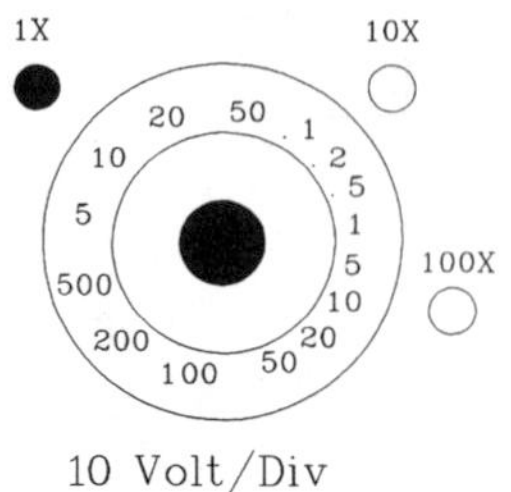

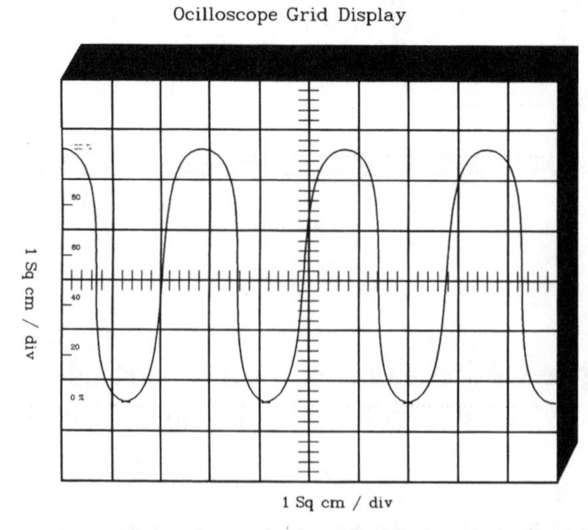

Figure 2-3. Display of waveform measuring 5 units vertically and a Vert/Div setting of 10 volts per division.

There are four common values used to measure continuously changing AC voltages on an oscilloscope. AC voltages can be measured in four electrical terms on an oscilloscope display. these are:

Peak to peak voltage
Peak voltage
Average voltage
Effective voltage or root mean square (RMS) voltage

When using a oscilloscope these voltage measurements become easy to calculate and/or view on the oscilloscope once the peak to peak voltage is known.

Note: Effective RMS voltage is only true when the signal is that of a pure sine wave; any deviation from a sine wave will result in a erroneous voltage reading.

To help understand AC values, lets discuss their meanings starting with the average voltage value. The *average voltage* describes the mean values of the waveform and is equal to waveform area divided by its length. The voltage term used most frequently in electronics, however, is the *effective voltage* or RMS. It is this electrical value that most electronic test equipment is designed to utilize. An example would be AC voltmeters and ammeters. They are designed and calibrated to read RMS.

DEFINING AC VOLTAGES

To help us more deeply define these common electrical terms as values lets work out an example. Example 3 will demonstrate the voltage calculations needed for each electrical term, and show each of their voltage relationships on the oscilloscope display as shown in **Figure 2-4**.

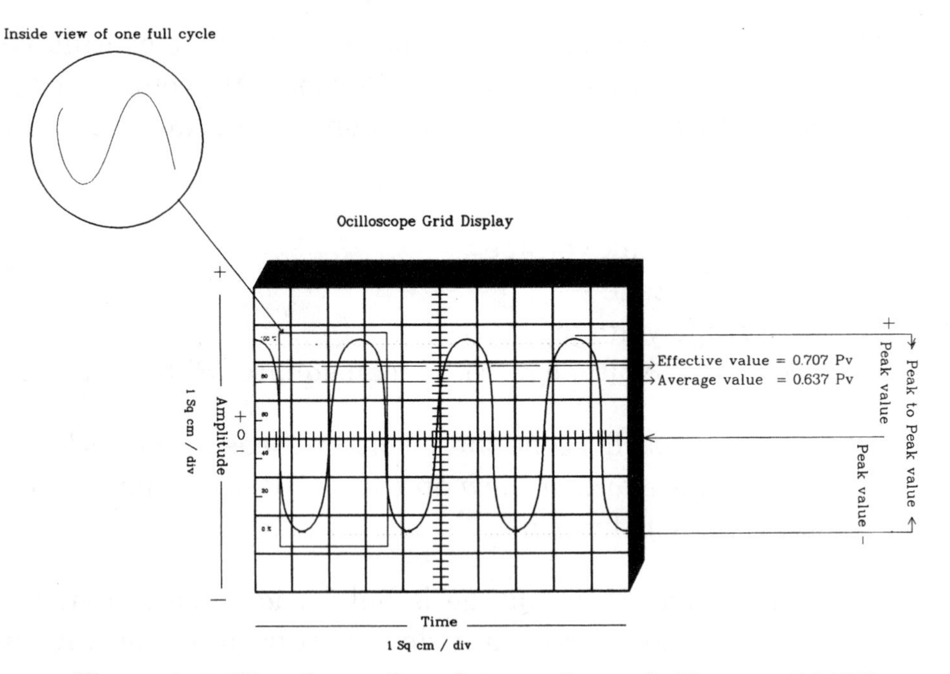

Figure 2-4. Waveform of peak to peak, peak, Vavg and RMS.

EXAMPLE 3:

The waveform in **Figure 2-3** measures 6 units vertically and the oscilloscope Volt/Div reads 10 volts per division. The signal voltage would be:

6 vertical units x 10 volts per division = 60 volts

What are the waveform voltages in example 3?

Peak to peak voltage (Vpp) = Vp/0.5 or 25Vp/0.5 = **50 Vpp**
Peak voltage (Vp) = 0.5 Vpp or 50Vpp 0.5 = **25 Vp**
Average voltage (Vavg) = 0.637 Pv or 25Vp 0.637 = **15.9 Vavg**
Effective voltage (RMS) = 0.707 Pv or 25Vp 0.707= **17.7 Vrms**

Note: All voltages are assumed to have a 0 volt ground.

As shown in Example 3, AC voltages can be measured in any of the four electrical terms when using an oscilloscope. Engineers and physicists who design electronic equipment using oscilloscopes need to know all the voltage terms, but the most important values that you will have to learn initially are peak to peak and effective.

MEASURING FREQUENCY AND PERIOD

A waveform that only has amplitude is of very little aid to the user an oscilloscope. A means must be provided to measure the waveform's frequency and period. The Tektronix 485 oscilloscope accomplishes this nicely by using the Time/Div control and the display grid. **Figure 2-5** shows the placement of the Time/Div control, and the oscilloscope display. The oscilloscope's Time/Div function controls the horizontal sweep, which is precisely calibrated in 27 ranges. See **Table 2-1** and **Figure 2-6**.

Sweep time is the speed that the beam moves from left to right on the CRT display. The sweep speed or rate is equal to the oscilloscope Time/Div readout multiplied by the number of X-axis squares and multiplied by the oscilloscope's unit of time (i.e., ms, μs, ns). Let's take a closer look in Example 4.

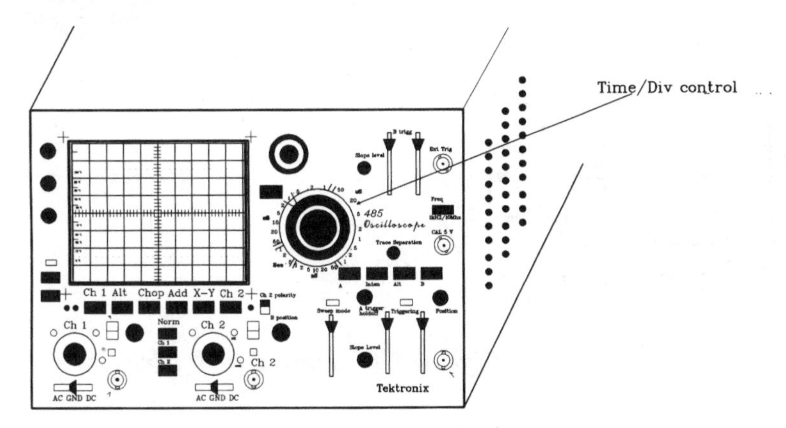

Figure 2-5. Drawing of Time/Div ranges.

Oscilloscope Time Ranges	Unit	Horizontal Sweep Rate
0.5, 0.2, and 0.1	seconds	
50, 20, 10, 5, 2, 1, 0.5, 0.2, 0.1	ms	x 10^{-3}
50, 20, 10, 5, 2, 1, 0.5, 0.2, 0.1	μs	x 10^{-6}
50,20,10,5,2,1	ns	x 10^{-9}

Table 2-1.

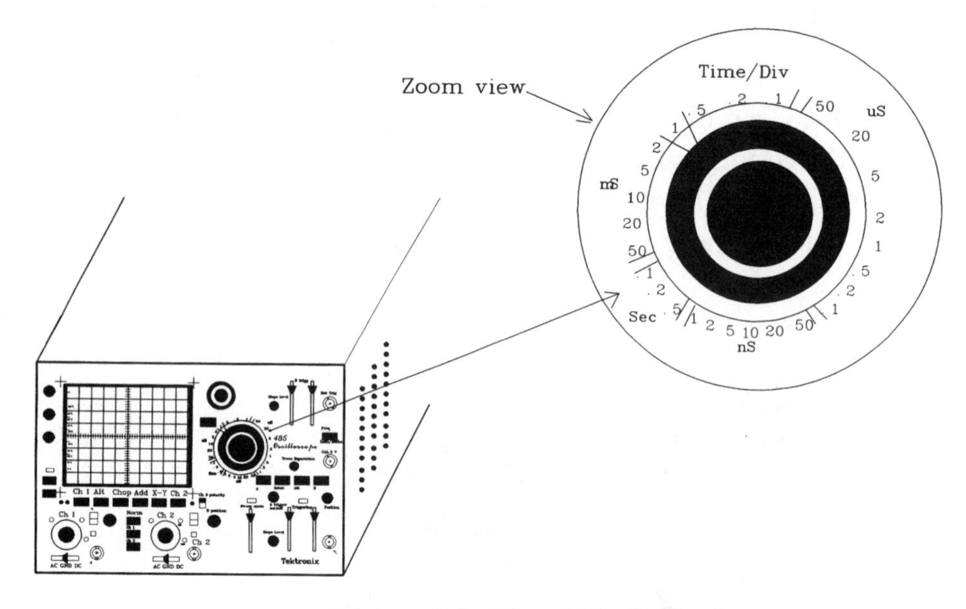

Figure 2-6. Time/Div & display.

EXAMPLE 4:

What is the oscilloscope sweep rate in seconds if the Time/Div is set to 5 ms, and the oscilloscope has 10 horizontal squares on the X-axis?

Sweep speed = (Time/Div) x (number of X-axis squares) x (power of ten)

Sweep speed = $5 \times 10 \times 10^{-3}$ = 0.05 seconds or 0.005 seconds per division.

Note: The number of ranges can vary depending on the oscilloscope's bandwidth and model.

If we are measuring an AC signal that is changing or repeating itself in a given time the law of physics states it has a frequency. Frequency is measured in hertz (Hz) and is equal to the number of times the signal repeats itself in one second. This is known as *cycles per second* (cps).

The repeating signal also has a period which is equal to the reciprocal of the frequency (i.e., one divided by frequency). The period of a signal is equal to the time the signal takes to complete one cycle.

The oscilloscope measures the period or the amount of time a signal takes to complete one cycle. By adjusting the Time/Div control the period measurement can be displayed on the X-axis as seen in **Figure 2-7**. Lets calculate the frequency using Example 5 and determine the waveform period in Example 6.

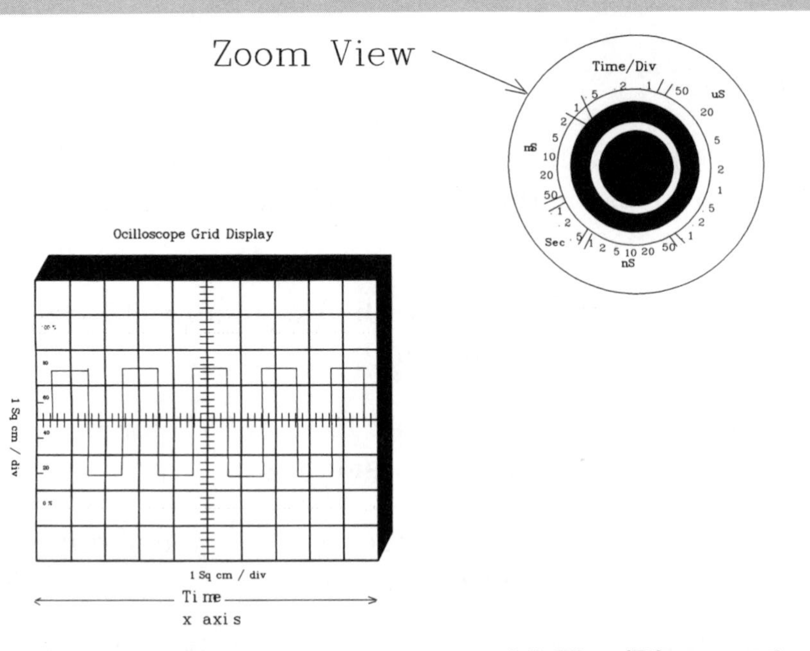

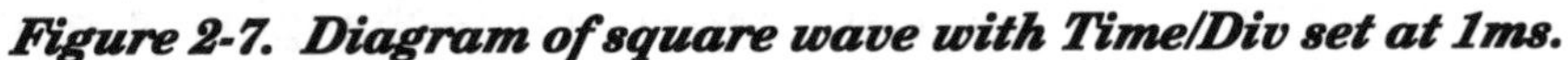

Figure 2-7. Diagram of square wave with Time/Div set at 1ms.

EXAMPLE 5:

Using **Figure 2-7** it can be determined that the square wave takes exactly one division to complete one cycle. Each division is equal to .001 second, per the Time/Div setting, so the frequency can be determined by taking the reciprocal of the period, or:

Frequency = 1/period
Frequency = 1/.001 = 1,000 Hz

EXAMPLE 6:

Using **Figure 2-8** the period can also be determined by taking the reciprocal of the frequency, or:

Period = 1/frequency
Period = 1/1,000 = .001 second

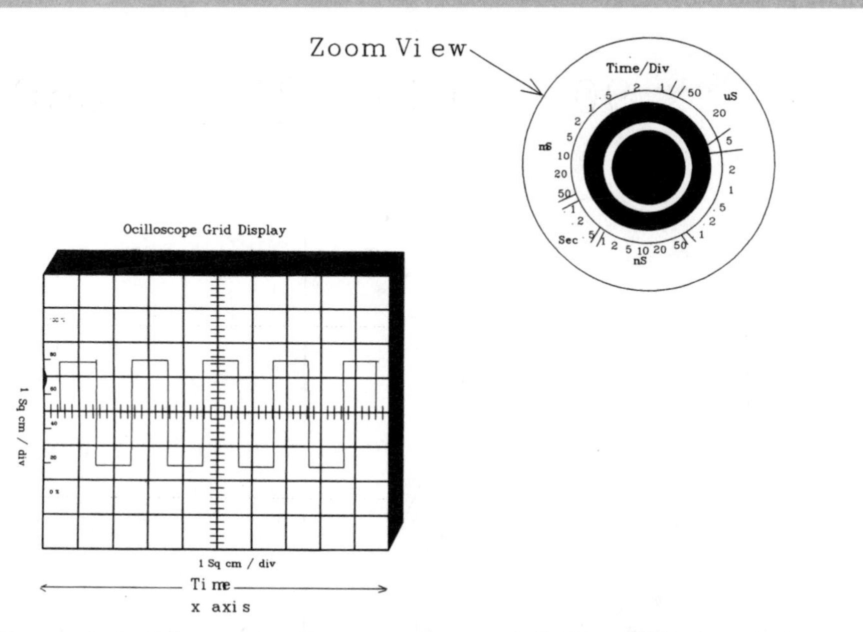

Figure 2-8. Diagram of square wave with Time/Div set at 5µs.

SIGNAL PHASE AND PHASE SHIFT

The phase of an AC signal can best be determined by viewing a sine wave on an oscilloscope and imagining one cycle of the sine wave as a complete circle consisting of 360 degrees (**Figure 2-9**). Using degrees you can refer to the sine wave's phase angle or determine the amount of waveform period that has changed over time, as shown **Figure 2-10**.

Phase shift is the difference in timing or degrees between two signals displayed simultaneously on the oscilloscope display. When comparing two similar signals the phase shift can be determined by measuring the difference of the first wave to that of the second wave by noting the start and finish times of the first wave and then comparing it to the start and finish times of the second wave. The difference between the timing of each

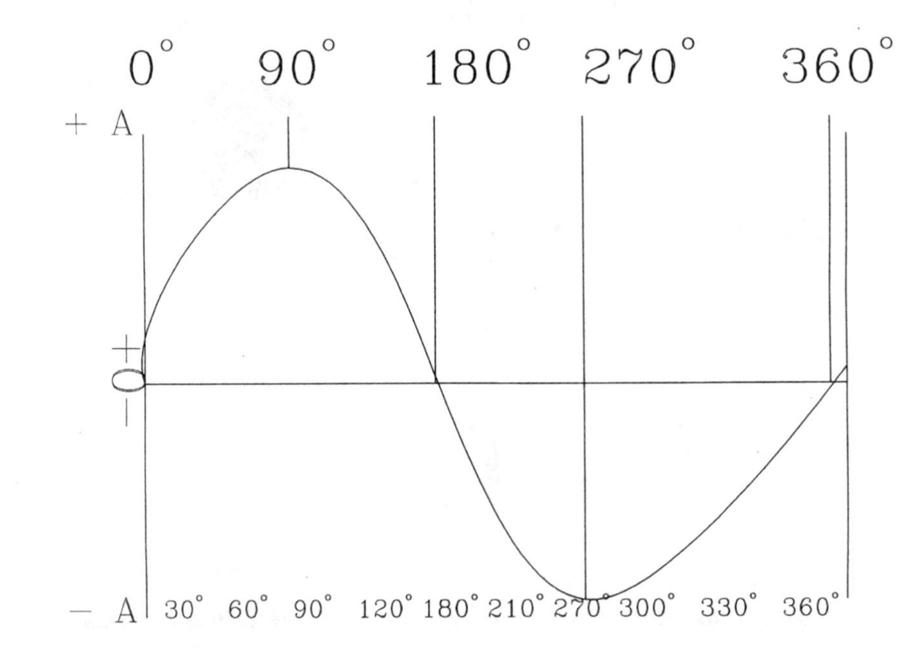

Figure 2-9. Simplified view of one cycle's angular equivalent on the CRT.

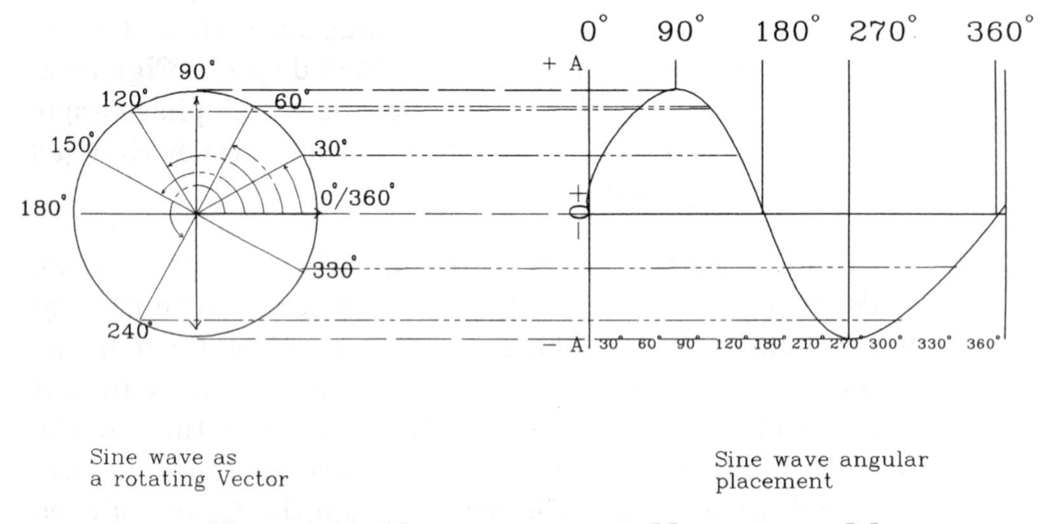

Figure 2-10. Sine wave generated by vector model.

wave is the phase shift, see **Figure 2-11**. Later in the book's chapters we will discuss an easier method to determine signal phase and phase shift.

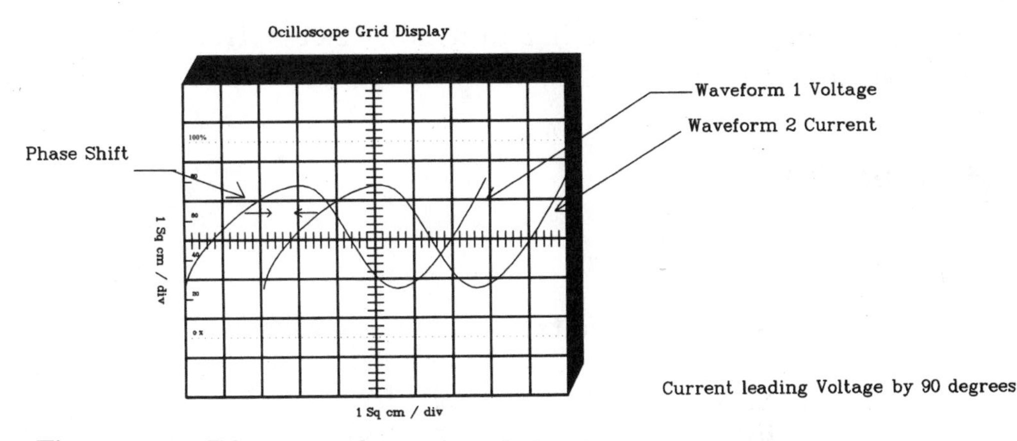

Figure 2-11. Diagram of two signals leading and lagging by 90 degrees.

SUMMARY

The oscilloscope display grid is made up from a number of square blocks. Each block is called a division and measures 1 sq. cm per division.

The oscilloscope display grid is used for both amplitude and frequency measurements.

The oscilloscope Volt/Div control is used for adjusting the signal level to the CRT.

The oscilloscope Time/Div control can be set to measure the period of a waveform.

A signal that constantly repeats itself has a frequency.

Frequency and period are reciprocals of each other.

Phase shift is the difference between two similar signals in degrees.

Frequency is measured in a unit called hertz (Hz) or cycles per second (cps).

CHAPTER 2 QUIZ

1. What is the RMS voltage if the peak voltage is 12V?

   A. 8.4V B. 5V
   C. 6V D. 18V

2. What is the frequency if the period is 1 μs?

   A. 10 MHz B. 1 MHz
   C. 100 MHz D. 1000 MHz

3. What is the period if the frequency is 1 MHz?

   A. 0.000001 s B. 0.00000001 s
   C. 0.00001 s D. 0.01 s

4. What is the highest frequency that can be displayed if the oscilloscope Time/Div control reads 50 µs?

   A. 20000 MHz B. 50000 kHz
   C. 20 kHz D. 500 kHz

5. AC voltages can be measured in how many terms?

   A. 6 B. 7
   C. 4 D. 13

6. Can a DC signal have a frequency?

   A. Yes B. Sometimes
   C. No D. No one knows

7. What determines the period of a waveform?

   A. Amplitude B. Gain
   C. Cycle time D. Voltage

8. What is the frequency if the period is 10 μs?

   A. 100 kHz B. 100 MHz
   C. 10 kHz D. 1000 kHz

9. Which is not a signal characteristic for proper display viewing?

   A. Phase shift B. Shift in phase
   C. Phase D. Differential quality

10. One complete cycle has how many degrees?

    A. 45 degrees B. 180 degrees
    C. 360 degrees D. 270 degrees

11. Measuring only the positive peak of a signal is the?

    A. Vpp B. Vp
    C. RMS D. Average voltage

12. If a signal measures 70.7 V and the peak voltage is 100 V what voltage term is being measured?

    A. RMS B. Vavg
    C. Vpp D. Vp

13. How many cycles are in 10 MHz?

    A. None B. 10 million
    C. 1 million D. 1 hundred

14. What axis of an oscilloscope is frequency measured on?

    A. X-axis B. Z-axis
    C. Y-axis D. None of the above

15. Frequency is measured in what unit?

    A. Volts
    B. Kelvin
    C. Amps
    D. Hertz

NOTES

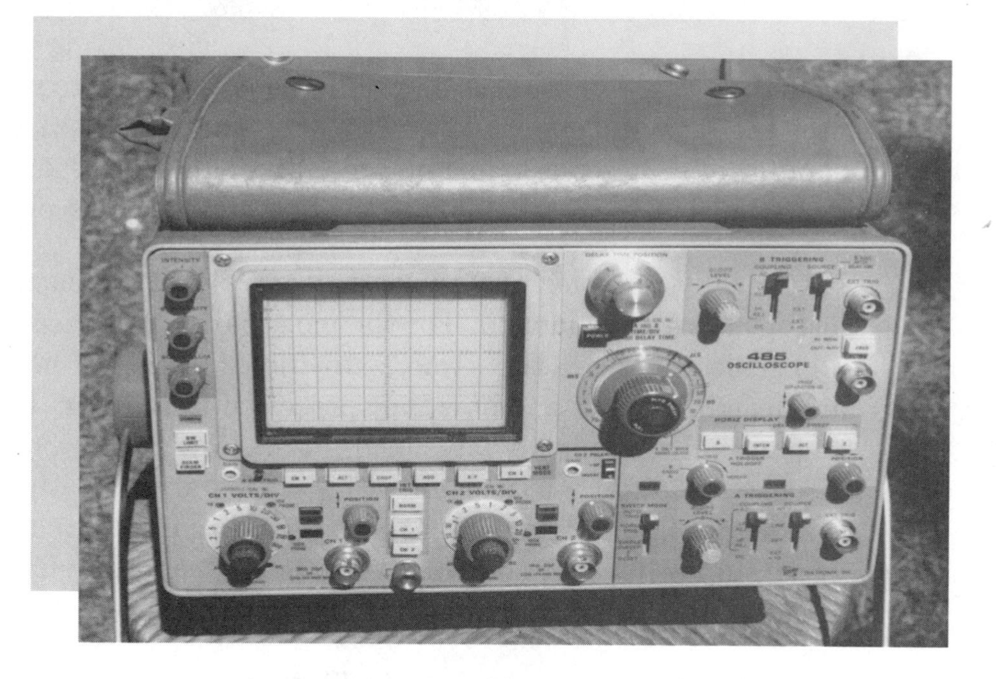

CHAPTER 3
OSCILLOSCOPE CONTROLS

CONTENTS:

Now that we know what is an oscilloscope, and how to in interpret the oscilloscope waveforms, let's take a moment to talk about the oscilloscope's controls, and functions that deserve special attention, especially among the first time oscilloscope user.

Historically, for over fifty years the controls on the basic oscilloscope have not changed. For the most part it can be said that all oscilloscopes are the same, with the exception of how they are designed today, and how they process information. They all do one thing (display a waveform.) Later in chapter 5 we will discuss the different types of oscilloscopes and how they process signal information. But for right know it is important to understand the basic controls of the oscilloscope, and how they affect all the pixels and qualities on the oscilloscope display.

The basic oscilloscope. as seen in **Figure 3-1**, consists of many front panel controls and some minor controls in the back of the unit. It is the author's goal to acquaint you with only the most rudimentary controls to get you started, but by no means are these to be considered all the controls of an oscilloscope, nor keep you from exploring all the controls on your own oscilloscope.

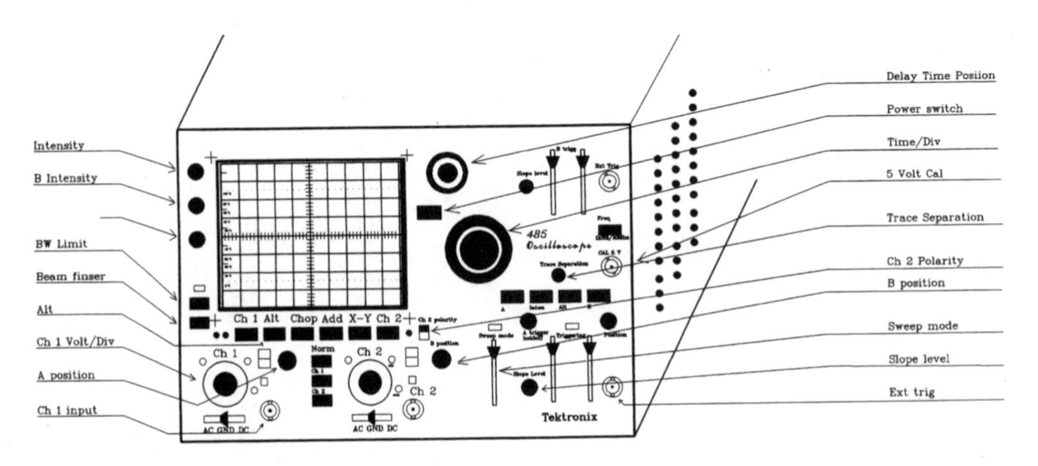

Figure 3-1. Picture of basic oscilloscope functions and controls.

When first viewing the front display of an oscilloscope's panel, one may be a little afraid or intimidated by all the controls of the oscilloscope, but if you can, close your eyes, and imagine you're adjusting a small television.

Note: Always refer to your oscilloscope's operators manual or users guide for help.

The main controls of oscilloscopes are almost like that of a TV. In fact the basic oscilloscope is a form of television, except for a few more controls and/or functions. With the proper adapter and/or interface an oscilloscope can display TV pictures, and TVs can display oscillographs.

An adapter in this sense is a *transducer;* a device which converts the electrical or mechanical characteristics of one application to that of another. An oscilloscope transducer converts the signal voltage of a test or particular application to that of the proper signal characteristics for displaying as a oscilloscope waveform. There are many different types of transducer used in electronics, science and physics. The one type of transducer we are all familiar with, and at times take for granted, is our ears. Our ears are a special type of transducer that converts mechanical energy (i.e., small changes in air pressure) to electrical energy which we then perceive as sound.

Nevertheless, before we talk about different oscilloscope applications let's first learn a few more oscilloscope basics.

CONTROL FUNCTIONS

Oscilloscopes come in all sizes and costs, from very small handheld units called portables to very large types found on the test benches of electronic shops. Analog type oscilloscopes are the most economical and most common among oscilloscopes, see **Figure 3-2**. Oscilloscopes consist of a number of basic functions and controls, however; for simplicity, the main controls are outlined in the following diagram of **Figure 3-3**.

Figure 3-2. 485 analog oscilloscope.

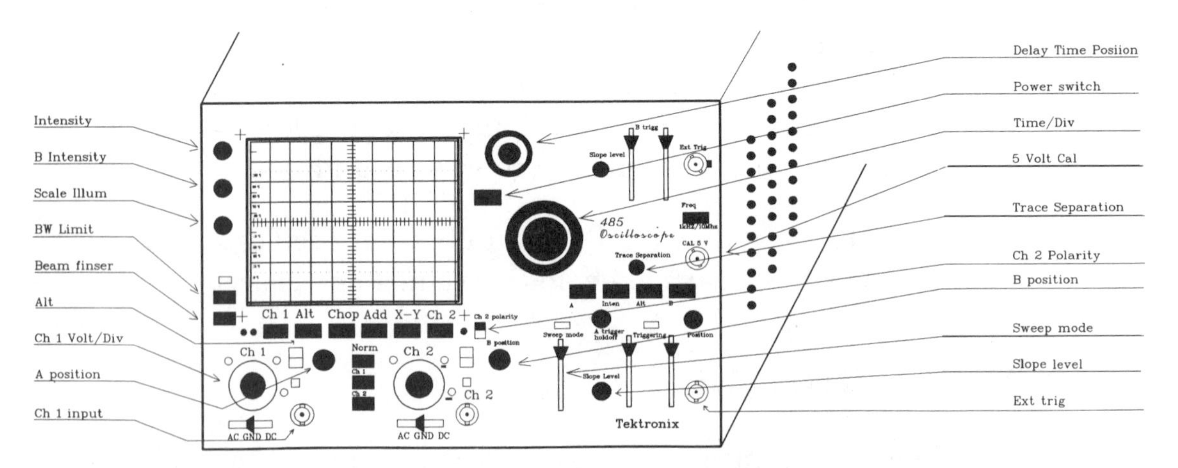

Figure 3-3. Diagram of oscilloscope's main controls.

Chapter 1 explained that the basic oscilloscope has five main functional sections: the vertical, horizontal, triggering, CRT and power supply. In the early years, oscilloscope manufacturers used clinical researchers to figure out what would be the most appealing to an oscilloscope user's sense of touch and vision. After vast amounts of research and user data had been gathered, manufacturers decided where to place the controls and what sizes they should be. In the late 1980's, oscilloscope manufacturers started to label the newer types of oscilloscopes in three defined sections on the front panel. The new oscilloscope panels were labeled *vertical*, *horizontal*, and *triggering*, and the oscilloscope display was tinted blue. This was a big change from the earlier models and the old green CRT display.

As more oscilloscopes were manufactured in the 80's, manufacturers and designers had seldom given any thought to the basic oscilloscope functions or user friendliness. Today, as oscilloscopes become more apparent in the communications and electronic service environments, it is of paramount importance to the first-time user to conveniently group the functional controls into categories according to order of operation, as shown in the oscilloscope diagram of **Figure 3-4**.

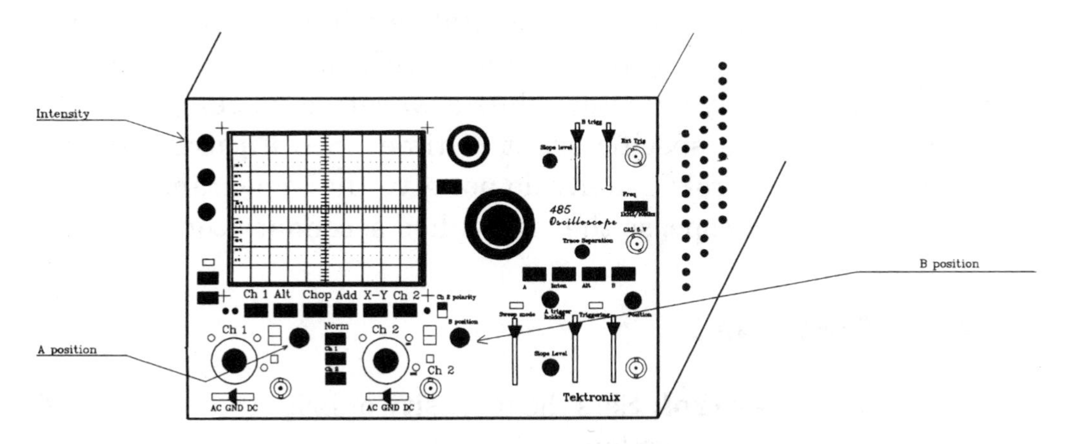

Figure 3-4. Picture of primary controls.

PRIMARY CONTROLS

As we view the oscilloscope diagram in **Figure 3-4**, the group of controls called primary controls (or functions) are:

Intensity control
Vertical position control (A position)
Horizontal Trace position controls (B position)

These three controls should be adjusted any time power has been applied to the oscilloscope. Failure to first make these adjustments could give the appearance of a defective oscilloscope. This is why they are called primary controls.

SECONDARY CONTROLS

The next group of the oscilloscope's controls called secondary controls (or functions). These include:

Amplitude setting (Volt/Div)
Horizontal time base (Time/Div)
Triggering

The secondary controls play a vital part in the enhancement of the viewing characteristics of the waveform. The secondary controls are usually adjusted while a waveform is being captured or viewing of a specific event in time. See **Figure 3-5**. The secondary controls usually do not give the appearance of a defective oscilloscope, so they are called secondary controls.

TERM DEFINITIONS

Intensity Control: Sets the intensity, or brightness of the CRT waveform on the display.

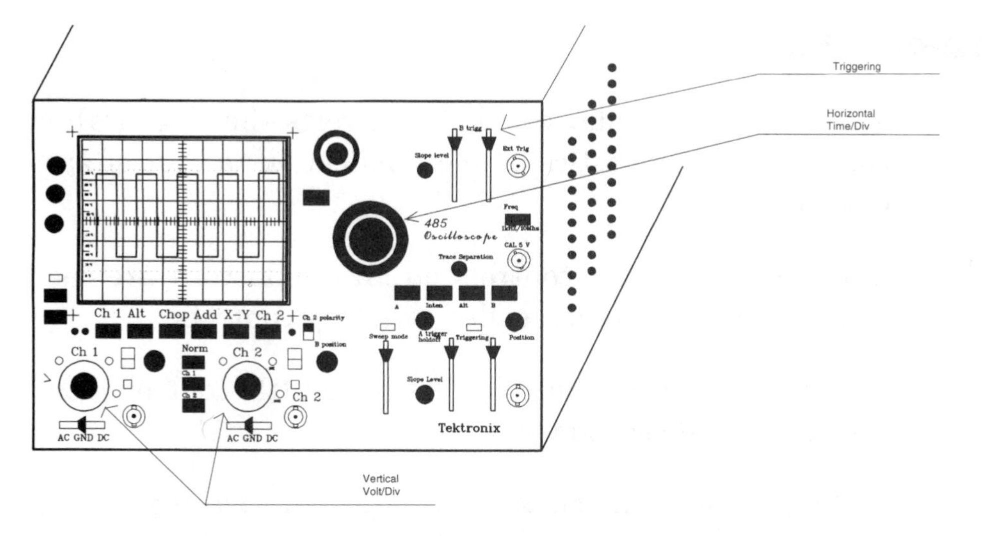

Figure 3-5. Oscilloscope secondary controls.

Vertical Position Control: Moves the Y-axis of the oscilloscope's trace on the CRT display in the up and down direction.

Horizontal Trace Position Control: Moves the X-axis of the oscilloscope's trace on the CRT display in the left or right direction.

Amplitude Ssetting: Controls the input signal via the Volt/Div control on the oscilloscope front panel. The Volt/Div adjusts the signal levels to the vertical amplifiers and CRT vertical deflection plates.

Horizontal Time Base: Controls the horizontal sweep time of the oscilloscope. The Time/Div control is used for waveform review and taking time and/or frequency measurements.

Triggering: Controls the waveform stabilities by freezing the picture for review. It functions in conjunction with the waveform type, horizontal sweep and vertical signal levels.

SUMMARY

A transducer is a device which converts the electrical or mechanical characteristics of one application to that of an other electrically.

The most important controls on an oscilloscope are the ones that control the picture, called primary controls.

The less important controls that have little control of the picture are called secondary controls.

If the primary controls are not adjusted correctly the oscilloscope will seem dead or defective.

All oscilloscopes, no matter how the controls are arranged, display waveforms.

If triggering is not utilized, the picture will not be stable.

All oscilloscopes have the same three sections in common: vertical, horizontal and triggering.

CHAPTER 3 QUIZ

1. What is the name of the square boxes on the CRT display?

   A. Graduals B. Divisions
   C. Graticules D. Cells

2. What is the shape of the CRT display?

   A. Round B. Rectangle
   C. Square D. None of the above

3. What control sets the CRT brightness?

A. Intensity
B. Light switch
C. Dim control
D. Purity

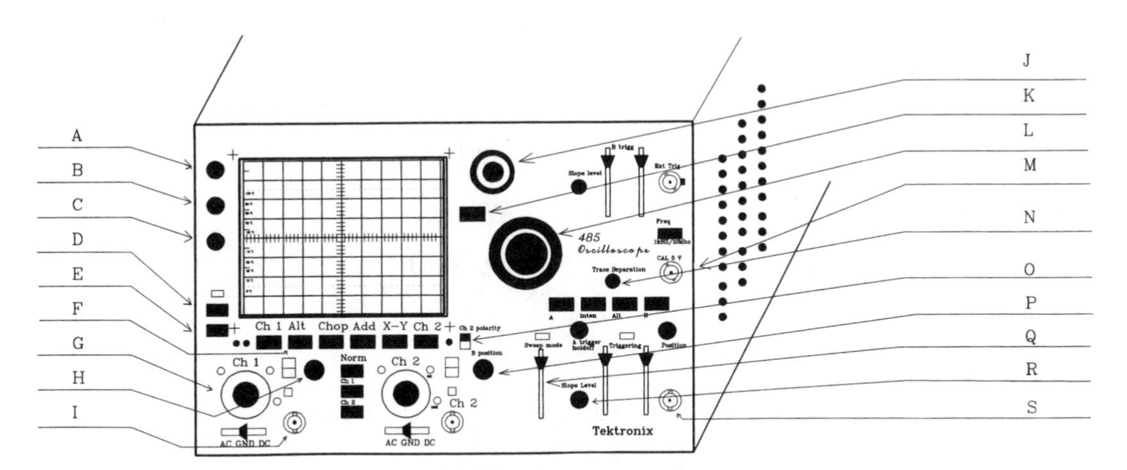

Figure 3-6. Oscilloscope for quiz.

4. How many squares are there on the X-axis of the CRT in **Figure 3-6** ?

A. 12
B. 6
C. 10
D. 2

5. Each square division in **Figure 3-6** measures ______?

A. 1 sq/cm
B. 1 sq/mil
C. 1 sq/in
D. 10 sq/cm

6. The biggest knob on an oscilloscope is usually the _______?

   A. Time/Div control
   B. Vertical control
   C. Trigger control
   D. Intensity control

7. If an oscilloscope appears to be inactive when first turned on what controls should be adjusted?

   A. Primary
   B. Secondary
   C. All
   D. None

8. What is the first control to adjust when using a scope?

   A. Intensity
   B. Focus
   C. Trigger
   D. Amplitude

9. What control is not considered a primary control?

   A. Vertical position
   B. Triggering
   C. Intensity
   D. Horizontal trace

10. What instrument has an aspect ratio of 1:1.2?

    A. Oscilloscope
    B. TV
    C. Eye glasses
    D. Computer screen

11. In **Figure 3-6**, what letter indicates the Time/Div control?

    A. C
    B. U
    C. L
    D. None of the above

12. In **Figure 3-6**, what letter indicates the Ch 1 input?

    A. P
    B. G
    C. I
    D. None of the above

13. In **Figure 3-6**, what letter indicates the trace position?

    A. H
    B. U
    C. C
    D. None of the above

14. In **Figure 3-6**, what letter indicates the Ch 1 Volt/Div control?

    A. G
    B. C
    C. X
    D. None of the above

15. In **Figure 3-6**, what letter indicates the trigger control?

    A. U
    B. P
    C. X
    D. None of the above

NOTES

CHAPTER 4
OSCILLOSCOPE USES

CONTENTS:

Oscilloscopes are used in many applications that require the technician or user to determine if a circuit or a particular function of a circuit is operating appropriately. Oscilloscope measurements assure the user that the test application is within its electrical specifications as designed.

This chapter will explain to you some of the many ways to employ an oscilloscope to test, or troubleshoot simple electronic devices and/or electronic equipment.

ANALOG CIRCUITS

In our first example let's use a single trace oscilloscope (i.e., one vertical input). In this example the oscilloscope will be used to view the output waveform of a 1 kHz square wave generator source connected to the input of an audio amplifier.

When the oscilloscope is connected as shown in **Figure 4-1**, a series of events start to happen. The oscilloscope displays a waveform that precisely follows the characteristics of the square

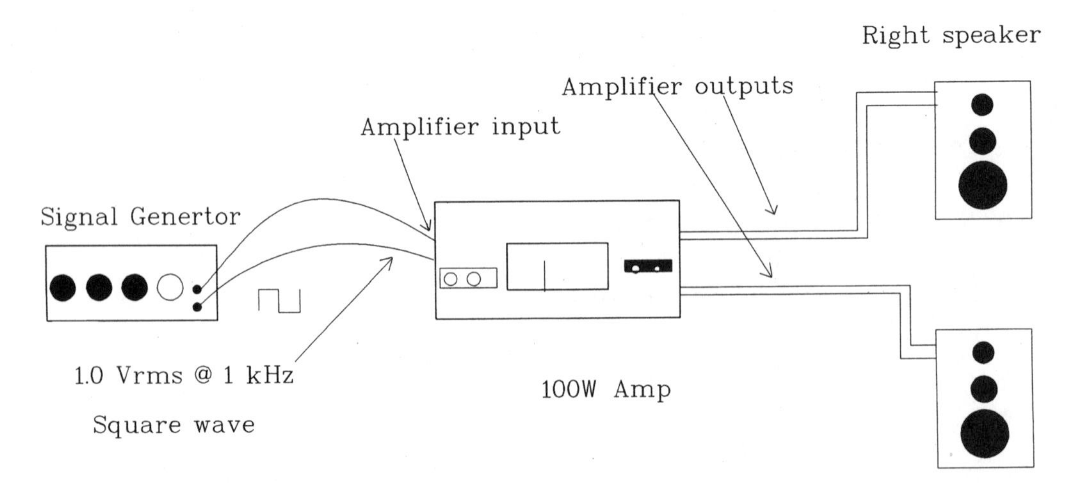

Figure 4-1. 1 kHz square wave connected to inputs of power amplifier.

wave generator and the audio amplifier output. The waveform will be instantly displayed in amplitude and time, as shown in **Figure 4-2**.

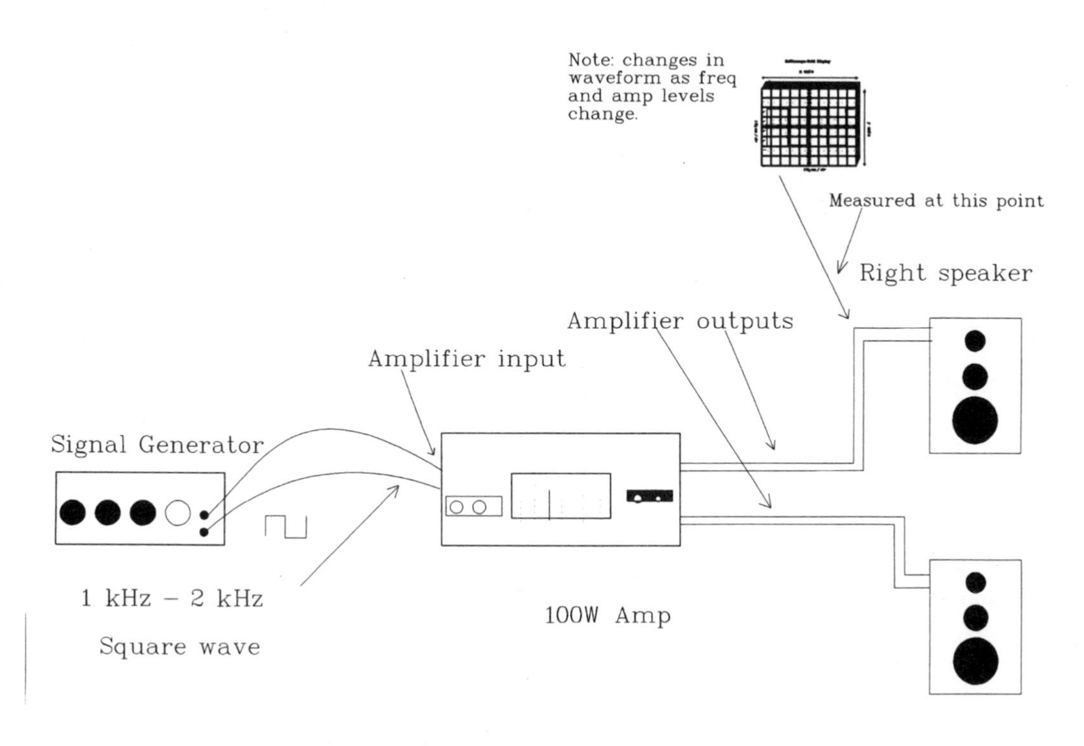

Figure 4-2. Waveform follows frequency changes & output levels of amplifier and generator.

By adjusting the horizontal time control the oscilloscope can display the square wave waveform as one full cycle occupying the full CRT screen as seen in **Figure 4-3**, or the user may desire to view a number of individual cycles of all equal time as seen in **Figure 4-4**.

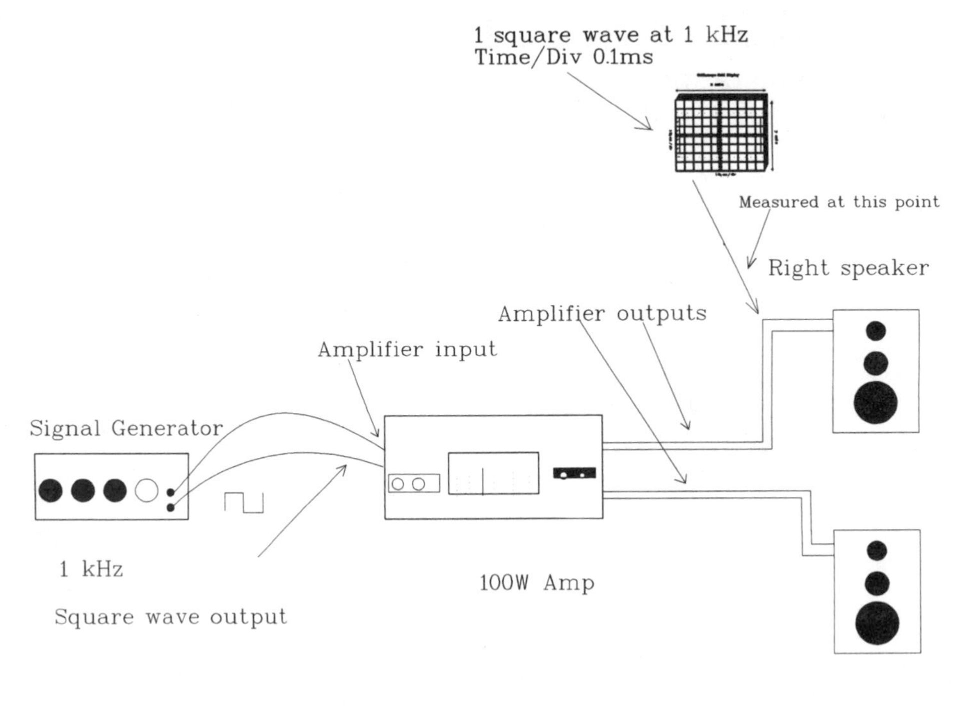

Figure 4-3. Waveform set to show 1 square wave at 1 kHz Time/Div at 0.1 ms.

When the oscilloscope is connected as show in **Figure 4-1** a low level signal from the generator is injected into the audio amplifier input. The technician can then adjust the amplifier controls and view on the CRT any noise and/or distortion characteristics of the audio amplifier. See **Figure 4-5**.

When servicing amplifiers the oscilloscope serves as an essential tool for measuring amplifier distortion and determining faulty components within the amplifier stages. Common problems in stereo amplifiers are when one amplifier channel is working or when both channels A and B are distorted.

It is the job of the oscilloscope user to interpret these waveforms and make visual determination on how he or she should proceed to the next logical step.

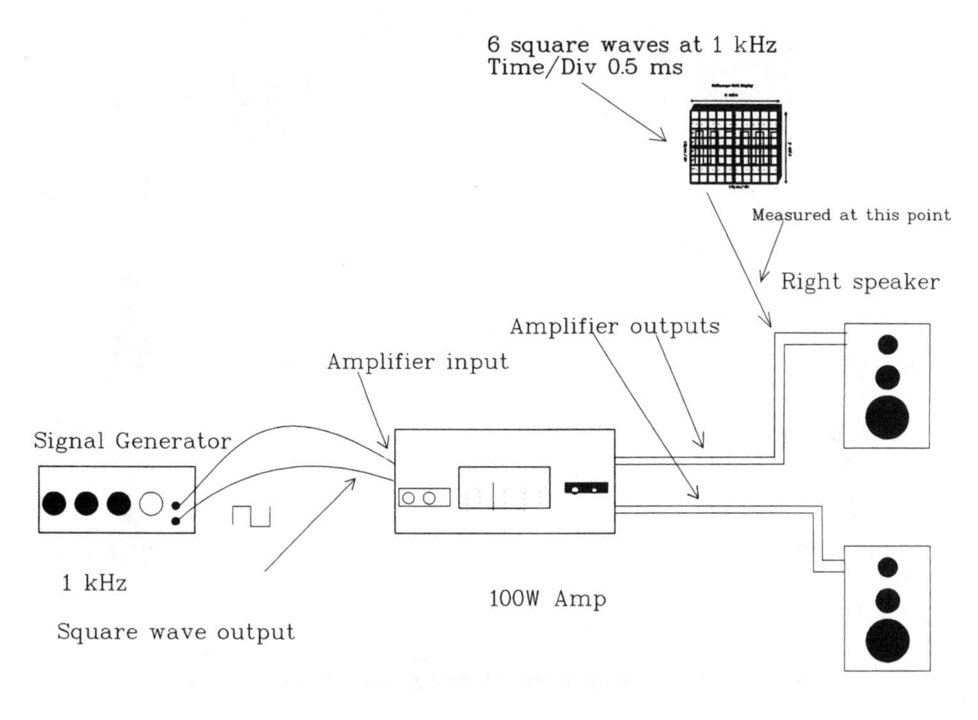

Figure 4-4. Waveform showing 6 square waves at 1kHz Time/Div at 0.5 ms.

In our next example, let's view an amplifier with output distortion using a 10 kHz square wave signal as a test source and a dual trace oscilloscope to display both amplifier channels A and B simultaneously. See **Figure 4-6**.

Note: Oscilloscopes with only one vertical input are called single trace oscilloscopes. Single trace oscilloscopes only display one waveform at a time on the CRT. Most high-end oscilloscopes are multi-traced, having a number of vertical inputs that can display a number of waveforms simultaneously.

This example shows that the output becomes distorted when we apply a signal to the input of the amplifier. If the user were to troubleshoot this amplifier he would simply use the oscilloscope to trace backward from the output stage until he or she

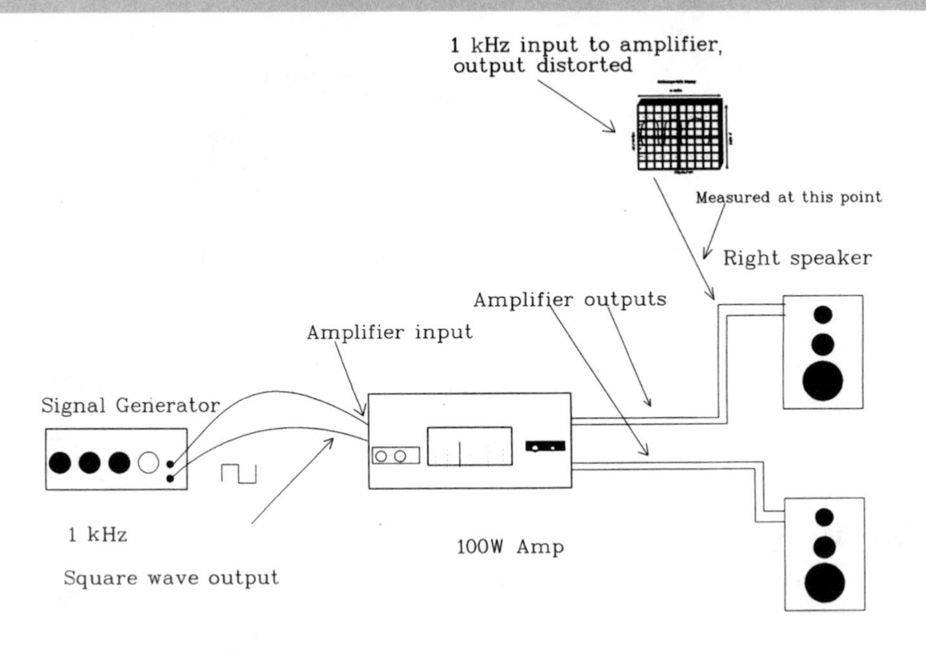

Figure 4-5. Waveform showing distorted output.

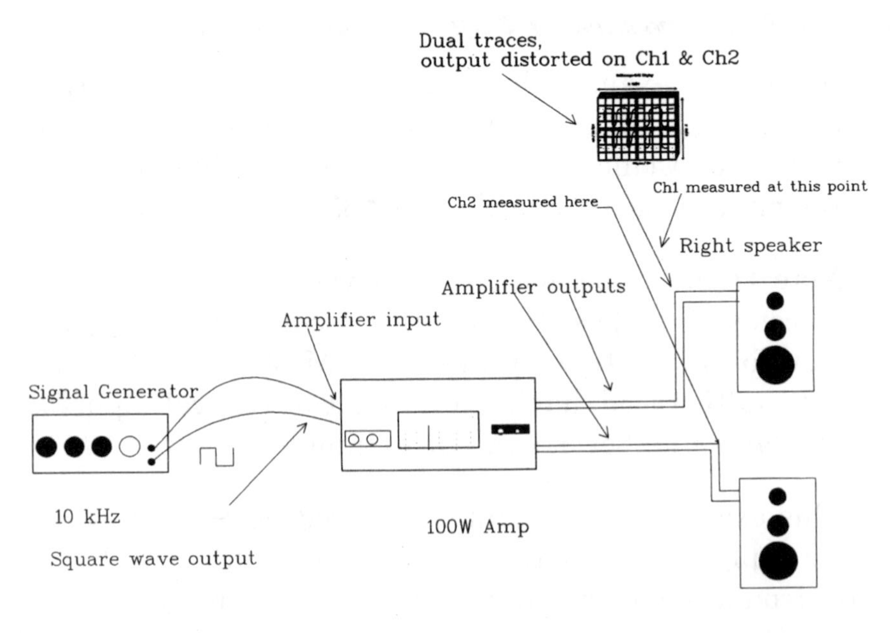

Figure 4-6. Dual trace waveforms with distorted outputs on Ch 1 and Ch 2.

saw a non-distorted square wave and then determine which component in the last stage was causing the distortion and replace it.

For simplicity, let's explore this simple testing technique using an amplifier block diagram, Figure 4-7, to point out each stage of the amplifier.

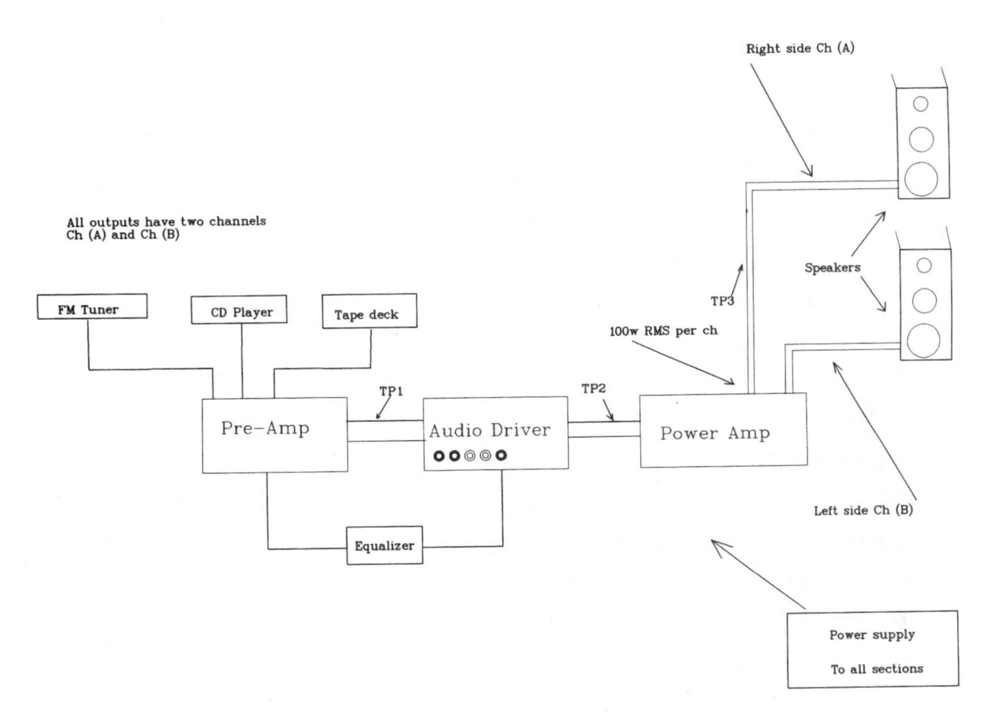

Figure 4-7. Block diagram of basic audio amplifier.

ACTIVE COMPONENTS

Electronic equipment consists a number of active and passive components. Active devices are all components that require power to become operational. For example:

Transistors
Diodes
Photo cells
Relays

Passive type components are the opposite of the above and usually do not require electrical power for their operation. For example:

Resistors
Capacitors

Usually the active components of electronic equipment are the first to become defective because of their switching characteristics, heat and/or power handling capabilities, creating current overloads, and/or distortion in the output stages of an amplifier.

Resistors and capacitors are not without blame, and resistors also become defective usually because of a direct result of an active component failure. Capacitors may become leaky or shorted when excessive current passes thought them, which can also cause other problems in electronic equipment. To help you understand how components fail, let's use the block diagram of **Figure 4-7** to help learn about troubleshooting techniques of electronic equipment.

As we probe or trace the amplifier diagram it becomes apparent that the output stage is defective. Why, because we see a clean signal at TP2 entering but a very distorted signal leaving the output at TP3 of the block diagram.

It now becomes the job of the oscilloscope user to determine which component is causing the problem. For the purpose of this example we troubleshooted this one for you and found a defective output transistor as the problem causing the distortion. The transistor was replaced and the amplifier was restored to its normal operation.

Note: Always remove the power when replacing defective components.

Noisy controls are another problem usually encountered in audio, video and communication equipment. A quick test via an oscilloscope can reveal the electrical characteristics of equipment controls and display their electrical performance. See **Figure 4-8**.

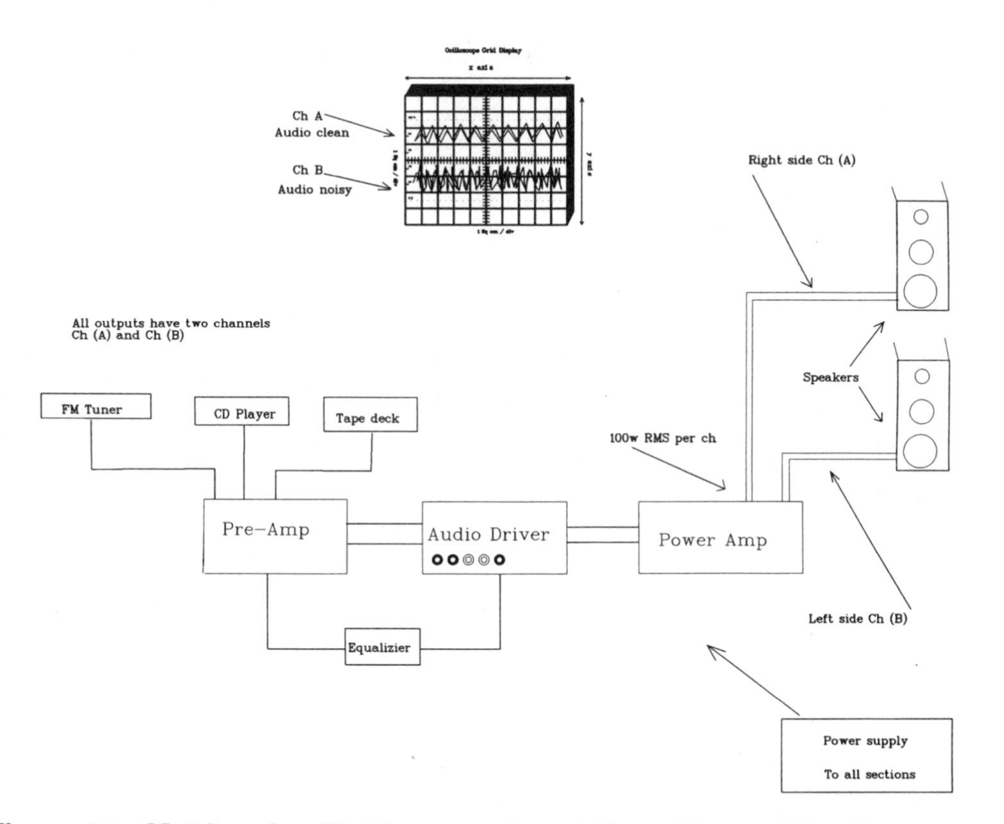

Figure 4-8. Ch 2 is noisy; Ch 1 has no noise while making audio adjustments.

The previous examples were all analog. They required no special equipment and were relatively simple. However, oscilloscopes are used in both analog and digital environments. Now that we have seen how we measure analog signals, let's take a few moments to explore digital measurements using an oscilloscope.

A digital signal consist of only 1's and 0's. They are sometimes called, HIGH or LOW or ON and OFF states. A digital signal can be likened to a light bulb being turned either off or on.

Oscilloscopes help the user or technician determine the state of digital equipment by showing the waveform patterns of the logic circuits in the forms of 1's and 0's, or high or low conditions of a digital circuit.

DIGITAL CIRCUITS

Digital input levels vary depending on the logic type (i.e., TTL, CMOS, etc.) and material in which the chip is made (i.e., positive logic or negative logic). In order not to confuse the reader we will confine our examples to only measurements showing High and Low states and encourage the reader to seek out more advanced books on the subject.

Measuring digital components is similar to making analog measurements with the exception that the technician is only concerned with the proper levels, and logic states (high or low) levels. Let's take a look at a few measurements using an oscilloscope to determine the proper logic output of different digital devices. See **Figure 4-9**.

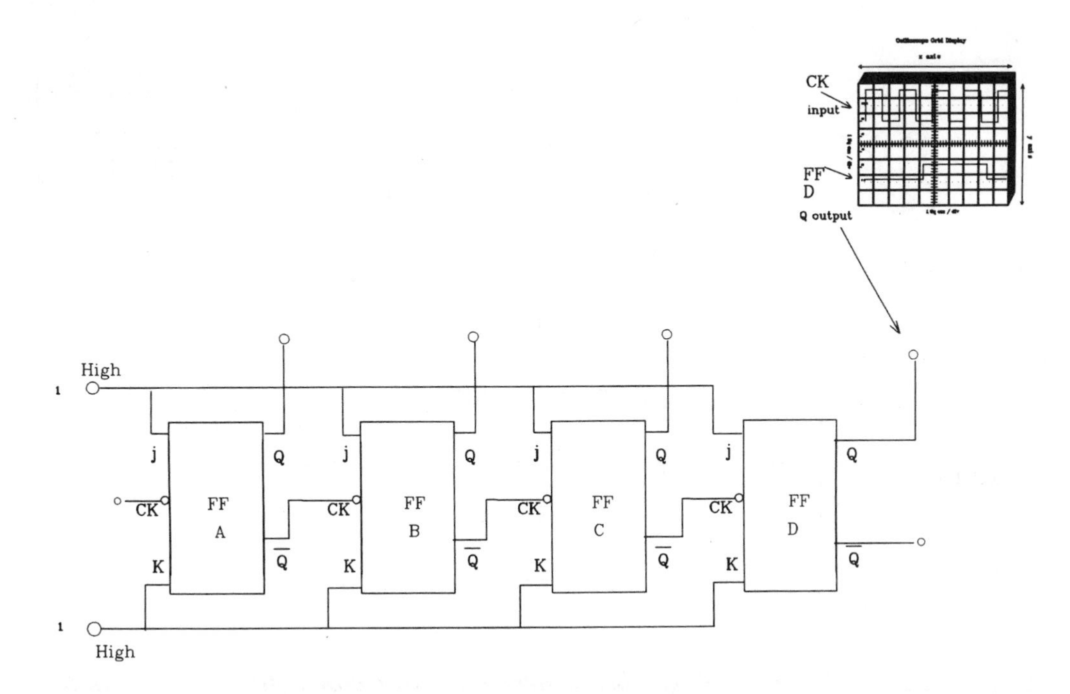

Figure 4-9. Diagram of logic levels on simple digital ripple counter.

The signal levels to the logic device inputs are critical to the device operation. If a logic device receives an input signal less or greater than it expects, the device will output a logic level that is not correct, see **Figure 4-10**. For this reason, digital circuits require a regulated power supply and good grounding techniques.

Note: The uses of oscilloscopes extend way beyond the scope of this book, and all interested readers are suggested to read more advanced literature on oscilloscopes used in science and physics to gain a wider perspective of its uses.

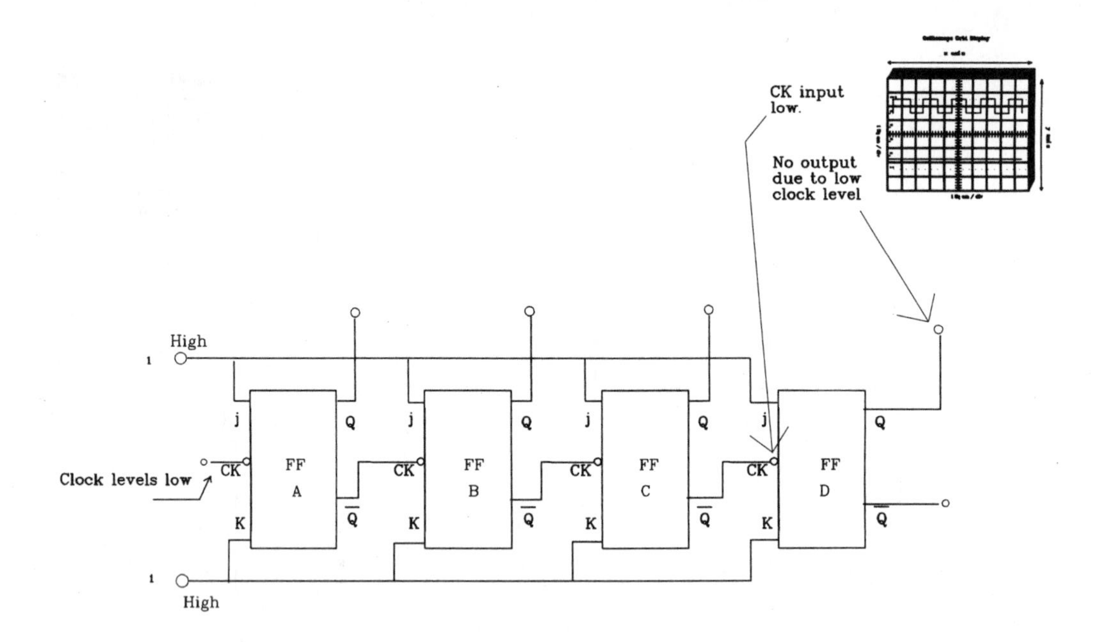

Figure 4-10. Diagram of ripple counter with low clock pulses causing no output on D flip-flop.

SUMMARY

An oscilloscope use the same controls for measuring an analog and digital signal.

An analog signal varies in time.

Active components usually become defective before passive components.

A transistor is an active device because it needs electrical power to operate.

Audio amplifiers are usually two channels but can have more.

Single trace oscilloscopes can only display one trace at a time.

Dual trace oscilloscope have two vertical inputs and can display two waveforms at the same time.

Digital signals only have two states (high or low).

Computers communicate in 1's and 0's only.

Digital circuits require regulated DC power for stable performance.

When replacing a defective component, first remove the power from the circuit.

Safety is always first when troubleshooting electronic equipment.

CHAPTER 4 QUIZ

1. A single trace oscilloscope has _______ vertical input(s).

   A. 2 B. 3
   C. 1 D. 4

2. Digital circuits require a(n) _______ power supply.

   A. Unregulated B. Pulse
   C. Ripple D. Sporatic

3. Computers communicate in _____ .

   A. 1's and 0's
   B. 2's and 4's
   C. 0's and 0's
   D. 1's and 1's

4. Which is not an active component?

   A. Resistor
   B. Transistor
   C. Diode
   D. Capacitor

5. A passive component _______?

   A. Needs power
   B. Needs no power
   C. Stays cold
   D. Overheats

6. The last components to fail in a circuit are _______?

   A. Passive
   B. RF
   C. Active
   D. Undetectable

7. A signal becomes distorted when a component _______?

   A. Shorts
   B. Opens
   C. Shorts or opens
   D. None of the above

8. A transistor is a ______ type of device.

   A. Passive B. Active
   C. Shorted D. None of the above

9. ______ is the primary cause of component failure in electronics?

   A. Shock B. Heat
   C. Poor construction D. None of the above

10. An AND gate has a logic high when both inputs are _____?

    A. Low B. High
    C. Neither D. Both

11. A logic gate which inverts the output level from the input level is a(n) _________?

    A. Inverter B. Emitter
    C. Ground Base D. None of the above

12. Grounding is more important in digital than _______ circuits.

    A. Analog B. VLF
    C. VLF and analog D. None of the above

13. An amplifier is defective if both channel outputs are ______?

    A. Distorted B. Shorted
    C. Open D. Closed

14. Digital devices have ______ state(s).

    A. 5 B. 2
    C. 1 D. 12

15. Safety is _____ when testing electronic equipment.

    A. Not important B. Number 1
    C. Impossible D. None of the above

NOTES

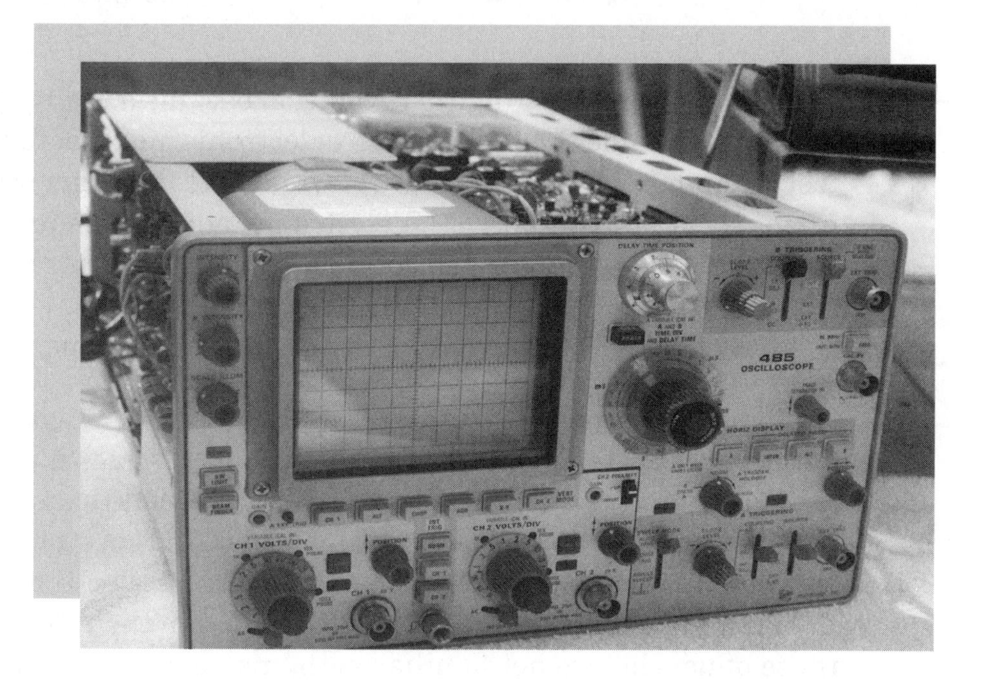

CHAPTER 5
ANALOG AND DIGITAL OSCILLOSCOPES

CONTENTS:

In the early 1980's the digital revolution became a reality. Ones and zeros became the topic of many electronic manufacturers of test equipment. Up until that time analog technology was the mainstay of test equipment for electronic industries, equipment manufacture and repair facilities. For this reason, this chapter will begin with the topic of the analog oscilloscope, and conclude with that of its successor, the digital oscilloscope.

Historically, years before the digital revolution technicians, scientists and those who had an interest in electronics used analog equipment. Educational institutions trained thousands of people using analog oscilloscopes and other types of analog equipment. Students became top engineers for electronic companies and technicians were in high demand in their fields. For those of us who are not familiar with analog, an example of an analog device would be a Volt-Ohm Meter, or playing an old 78 wax record on a turntable.

The simplicity of the analog oscilloscope made it an attractive test tool to many users, primarily because of its low cost and ease of operation.

Note: The position and size of each control on an analog oscilloscope is carefully thought out by a team of clinical researchers.

The basic analog oscilloscope consists of five electrical sections:

Vertical Amplifier
Horizontal Amplifier
Cathode Ray Tube (CRT)
Triggering
Power Supply

Figure 5-1 is a functional block diagram of an analog oscilloscope, excluding the power supply section.

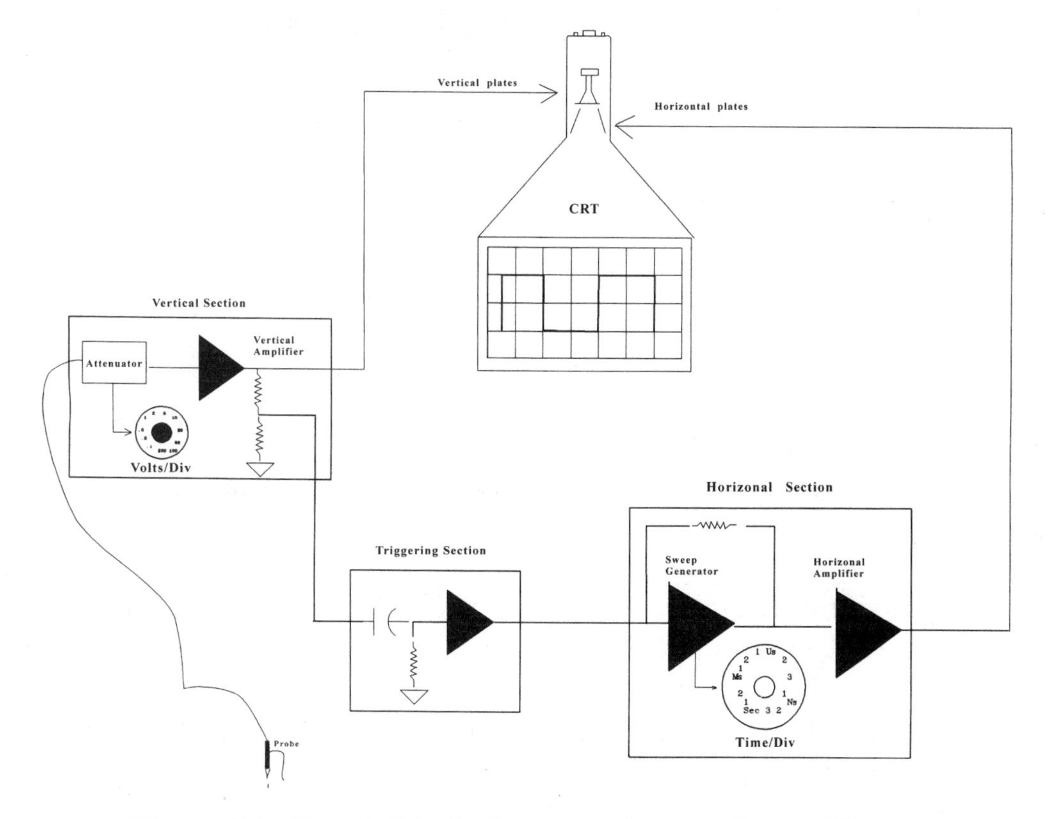

Figure 5-1. Simple block diagram of an analog oscilloscope.

ANALOG BASICS

The analog oscilloscope will function when a signal of proper amplitude and frequency is induced across the oscilloscope's vertical amplifier input. This is normally accomplished with the aid of two specially designed wires, called a *probe*.

The probe consists of a ground lead and a signal lead soldered to the tip of the probe point, or may be in a clip form for attaching to the circuit under test. See **Figure 5-2**.

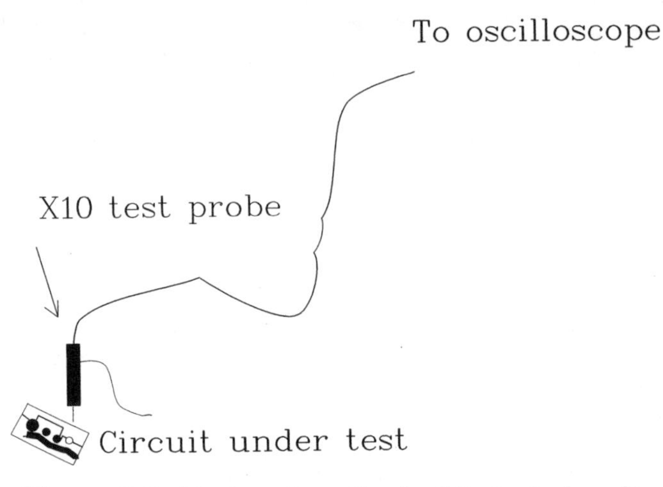

Figure 5-2. Test probe attached to test circuit.

The signal travels through the probe to the vertical amplifier where it is manually attenuated via the Volts/Div control setting by the user, and amplified to the proper level. Next, the signal voltage is fed to the CRT circuitry for vertical defection, as seen in **Figure 5-3**.

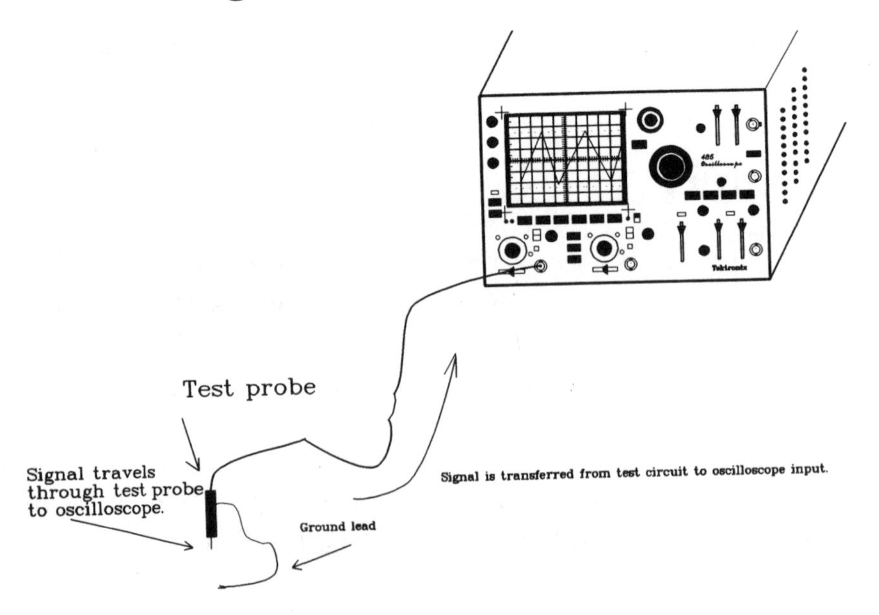

Figure 5-3. Test signal traveling from test probe to oscilloscope.

The cathode ray tube was developed in 1948 at Manchester University in England by professor Williams, and is the heart of the oscilloscope.

The function of the CRT is to convert electrical signals into a visual display. The tube contains an electron gun structured to provide a narrow beam of electrons which strike a phosphor screen which the electrons strike causing a light to be emitted in a small area or dot on the CRT screen in proportion to the intensity of the electron beam. See **Figure 5-4**.

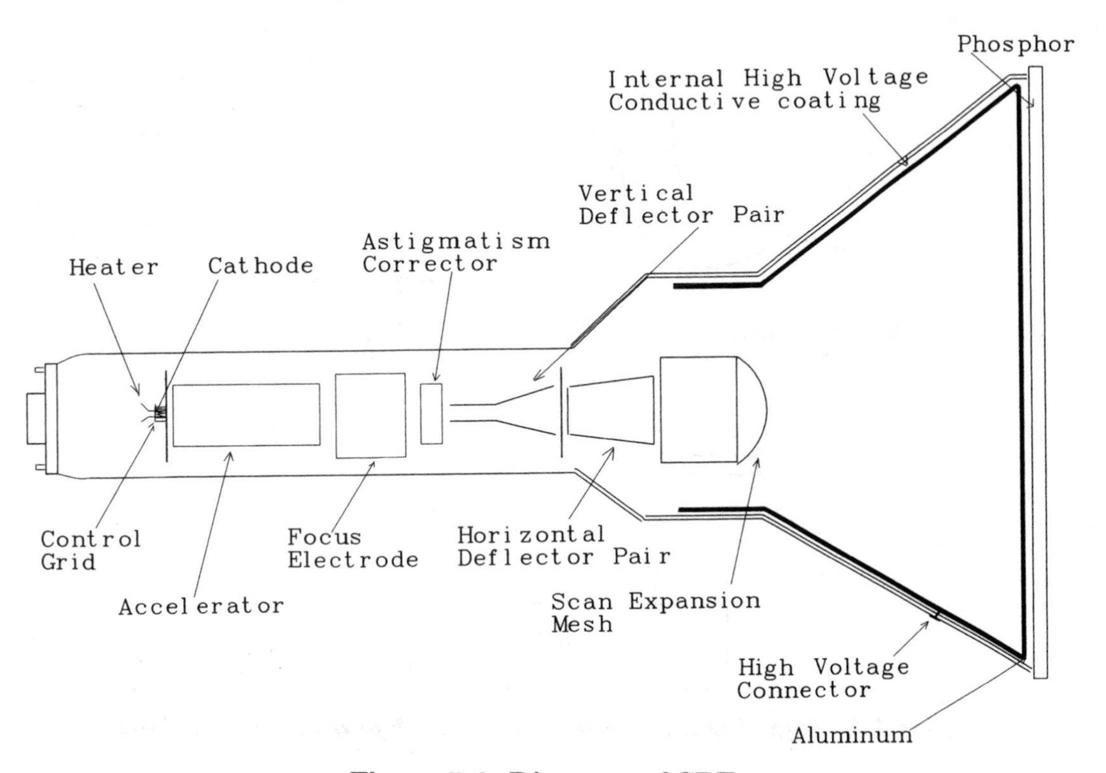

Figure 5-4. Diagram of CRT.

The test signal voltage from the test application travels through the oscilloscope probe to the vertical amplifier. See **Figure 5-5**. The vertical amplifier plays a vital role in the oscilloscope's ability to accurately display the rapidly varying signal.

Note: It is the vertical amplifier that determines the bandwidth characteristics of an analog oscilloscope.

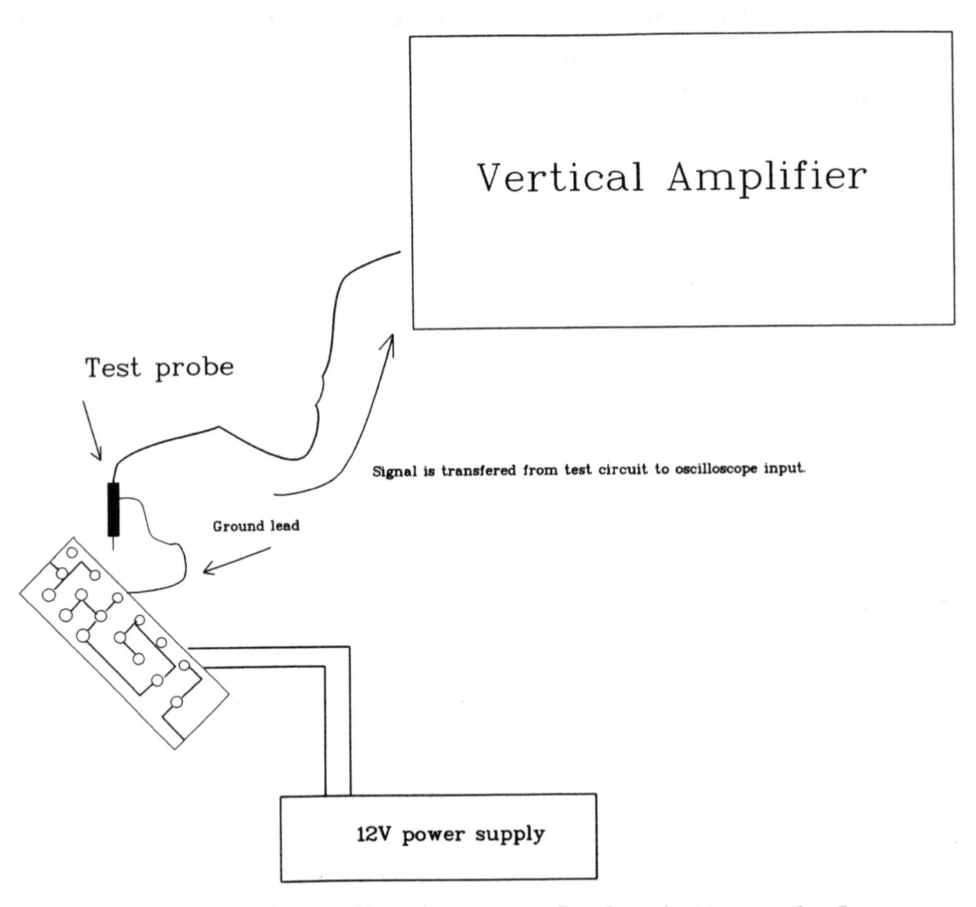

Figure 5-5. Signal traveling from test ckeck point to vertical amp.

Of equal importance, the vertical amplifier must not become saturated by the signal voltage or be used in a floating arrangement in a testing environment. Once the signal voltage has been properly attenuated and/or amplified, the signal voltage is sent to the vertical deflection plates inside the CRT. See **Figure 5-6**.

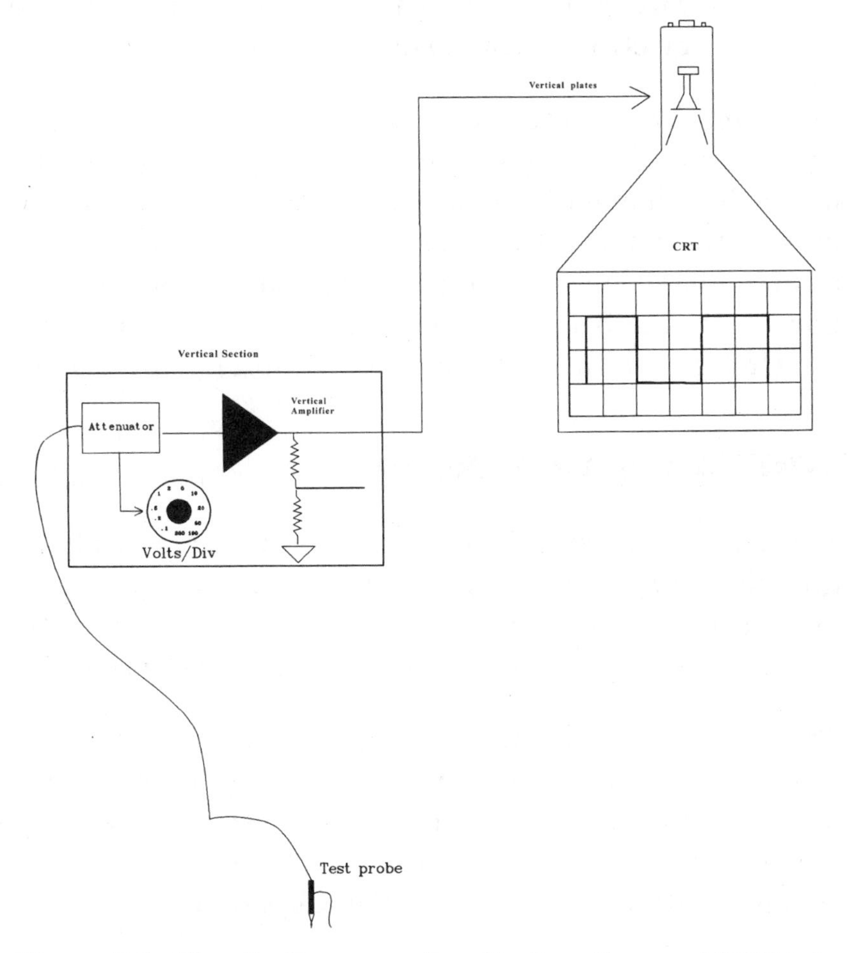

Figure 5-6. Simple diagram of vertical section and CRT tube.

It is the CRT that creates the small dot on the display which appears stationary until the signal voltage from the vertical amplifier and associated circuits causes the dot to move up or down. The deflection of the beam which causes the dot to move up and down on the CRT is proportional to the voltage applied to the vertical electrode within the CRT. The polarity of the signal voltage influences which vertical direction the beam moves. If the signal voltage is positive the dot or electron beam moves upward, and if the signal voltage changes to a negative value, the electron beam moves downward.

However, if we wanted to view a time dependent waveform in this manner the oscilloscope would be of little practical use to us. Although one dot is not enough to produce a waveform, by moving the dot over the entire CRT in a systematic manner the complete signal waveform can be reproduced. So, we must somehow cause the beam of electrons to move horizontally across the CRT screen in a manner like a graph is drawn on paper.

HORIZONTAL SWEEP FUNCTION

The horizontal sweep function moves the beam in a left to right direction in a fraction of time, called the sweep time or Time/Div control. The analog oscilloscope uses this function the same way as a digital oscilloscope would to measure time or frequency.

The digital oscilloscope, though, uses its time base and many other functions the same way an analog oscilloscope does. There are advantages to using analog type oscilloscopes, especially when measuring very fast waveform changes or making high frequency RF measurements. See **Figure 5-7**.

DIGITAL BASICS

The digital oscilloscope functions just like an analog oscilloscope, in the sense that they both display waveforms; however, the digital oscilloscope gathers the signal information differently.

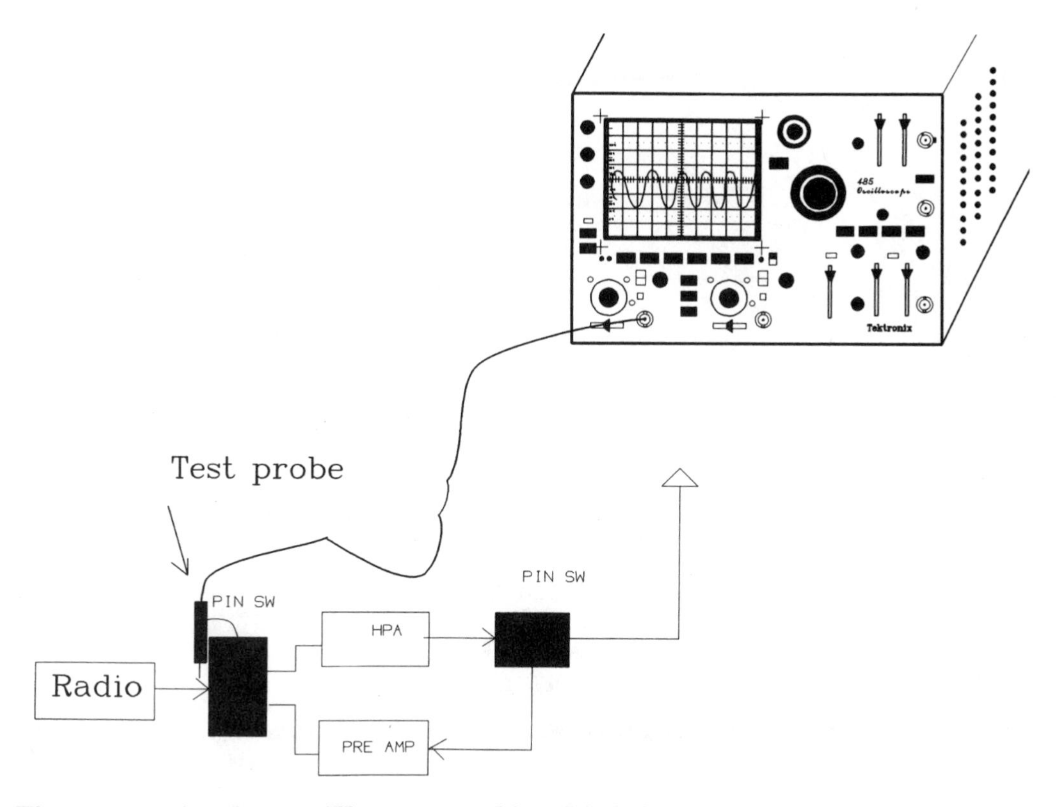

Figure 5-7. Analog oscilloscope making high frequency measurements on a communication system.

It does this by a unique means called *sampling*. The digital oscilloscope samples the signal voltage many times per second by a device called an *analog to digital converter* (ADC). **Figure 5-8** shows a digital oscilloscope block diagram.

The digital oscilloscope horizontal circuitry determines the number of samples the ADC takes and is measured in units called samples per second. The control of data acquisition and oscilloscope circuits that display the information are usually under control of a small computer called an "MPU" or application-specific ICs (ASICs) technology.

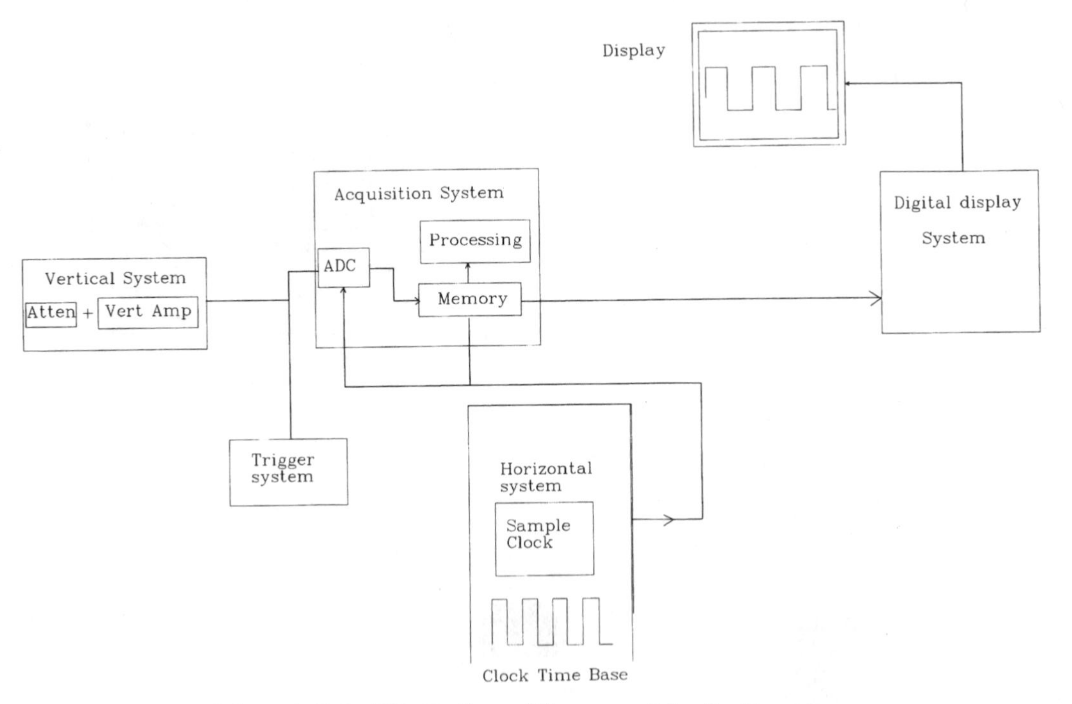

Figure 5-8. Digital oscilloscope block diagram.

Note: As of this writing Tektronix Corp. only produces digital oscilloscopes with a new technology called *InstaVu*.

The ADC converts the analog voltage from the vertical amplifier into a series of binary codes of eight or sixteen bit words. The ADC functions in real time which follows the input signal's amplitude and signal characteristics. See **Figure 5-9**.

When enough samples are taken and processed by the acquisition system, the data is sent to a digital to analog converter (DAC). The DAC converts the binary information into an analog signal which is displayed on the CRT as a waveform. **Figure 5-10** shows a photo of Tektronix's Digital Oscilloscope TDS 220.

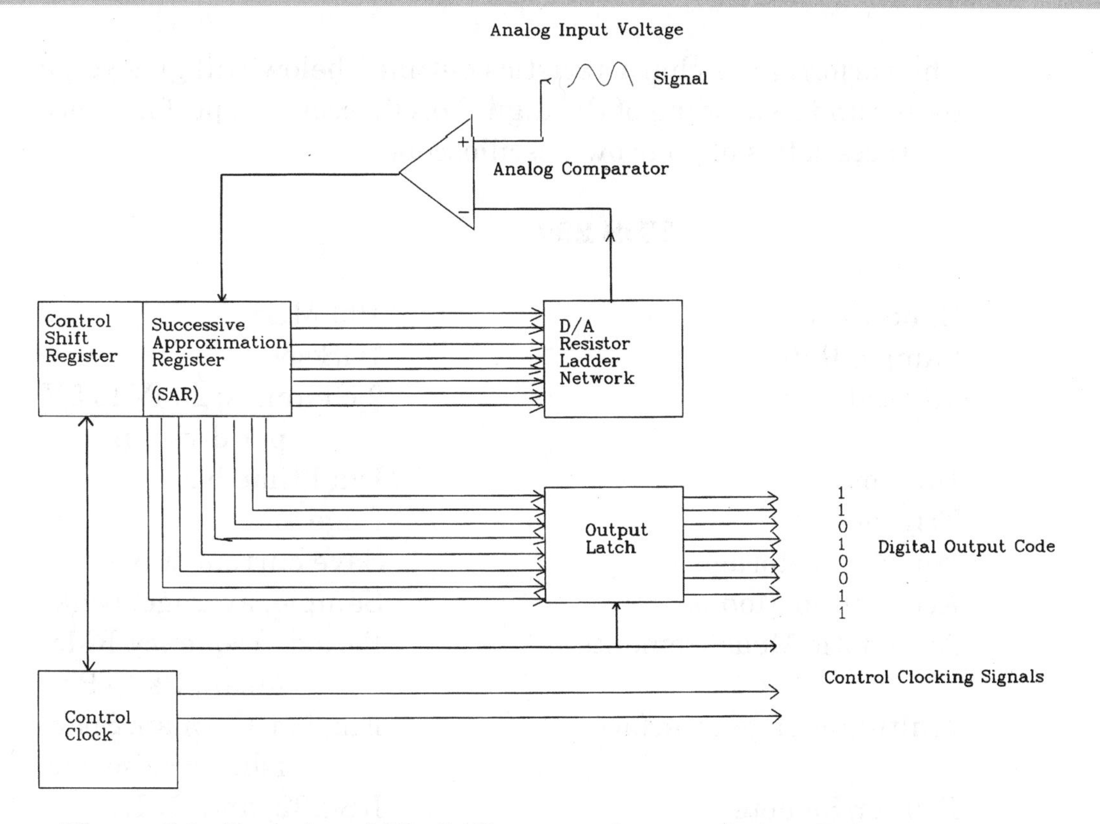

Figure 5-9. Functional block diagram of an analog to digital converter.

Figure 5-10. Photo of TDS 220 oscilloscope.

The performance characteristics outlined below will give you a better understanding of the digital oscilloscope vs. performance characteristics of your own oscilloscope.

TDS 220

Bandwidth	100 MHz
Sample Rate	1GS/sec
Vertical	2 channels; 2 mV to 5V per division
Horizontal	Dual time base
Trigger	Edge
Waveform storage	Save current wave
Acquisition Modes	Sample, average, peak
Automatic Measurements	Period, frequency, RMS, Mean, Pk to Pk
Multi-language interface	English, German, Spanish, Chinese, etc.
Printer/Remote	RS-232, and GPIB

INFORMATION PROCESSING

The speed at which information can be processed highly depends on the sampling rate and how the data is collected by the oscilloscope. Most digital oscilloscopes collect data in a single pass or in real-time sampling, but when the signal is very fast the digital oscilloscope can not collect enough samples to reconstruct an accurate waveform. In this high speed case, digital oscilloscopes use what is called *interpolation.*

Interpolation is a mathematical processing technique which entails taking a given set of points and fitting a function to them. In other words, with the help of interpolation the oscilloscope can estimate what the waveform should look like based on a mathematical function. Interpolation can build a high-speed waveform over time as long as the signal repeats itself; this is called *equivalent time sampling.*

However, a better form of sampling exists and is more versatile in its approach then equivalent time sampling. It is called *digital real time* (DRT) sampling. DRT samples at a rate several times that of the input frequency, acquiring enough points from each cycle to reconstruct it faithfully. DRT captures the entire signal in one trigger event and accurately captures irregular signals.

There are many types of sampling theorems used in digital signal processing (DSP). DRT interpolation used in digital oscilloscopes is only one type of sampling theorem used, other include:

Nonuniform sampling
Kramer's generalization
Papoulis' generalization
Lagrangian interpolation
Continuous sampling
Linear interpolation
Trigonometric interpolation

Interestingly, of these seven theorem types, the oscilloscope designers use only two which deserve limited explaining.

Digital oscilloscopes use real time sampling as their standard for measuring signals; however, as the signal becomes faster and faster the digital oscilloscope cannot collect enough samples to reconstruct the waveform. Only a few samples are ever collected, but of this few the waveform can be reconstructed by using linear interpolation and/or trigonometric interpolation.

Linear interpolation, as seen in **Figure 5-11**, reconstructs the waveform with a series of lines. Trigonometric interpolation reconstructs the waveform with a series of curves, as seen in **Figure 5-12**, and accurately displays the waveform with only a few samples taken.

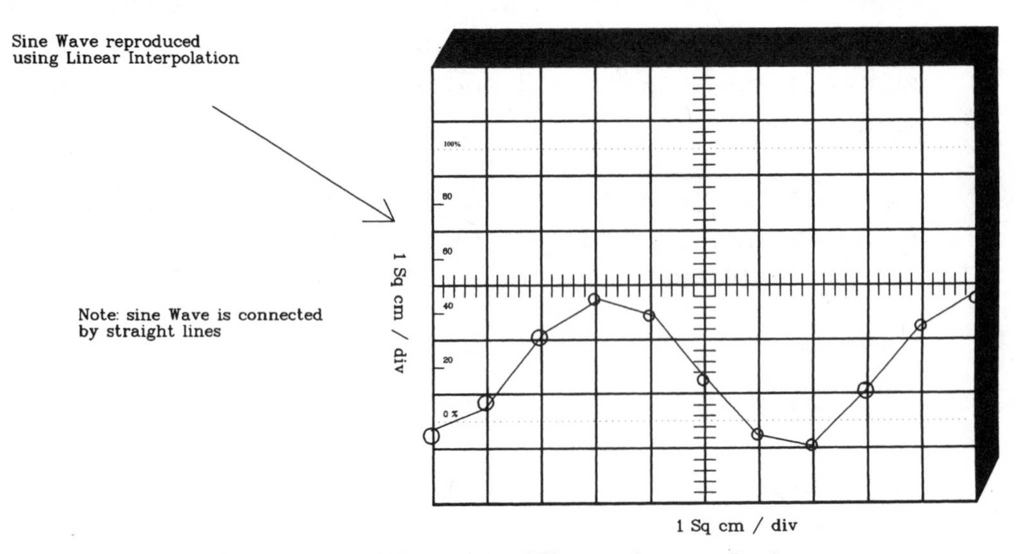

Figure 5-11. Diagram of linear interpolation.

SUMMARY

Analog oscilloscope are mainly used for fast and very high-frequency viewing.

Analog oscilloscopes are simpler in design.

Analog oscilloscopes cost less than digital oscilloscopes.

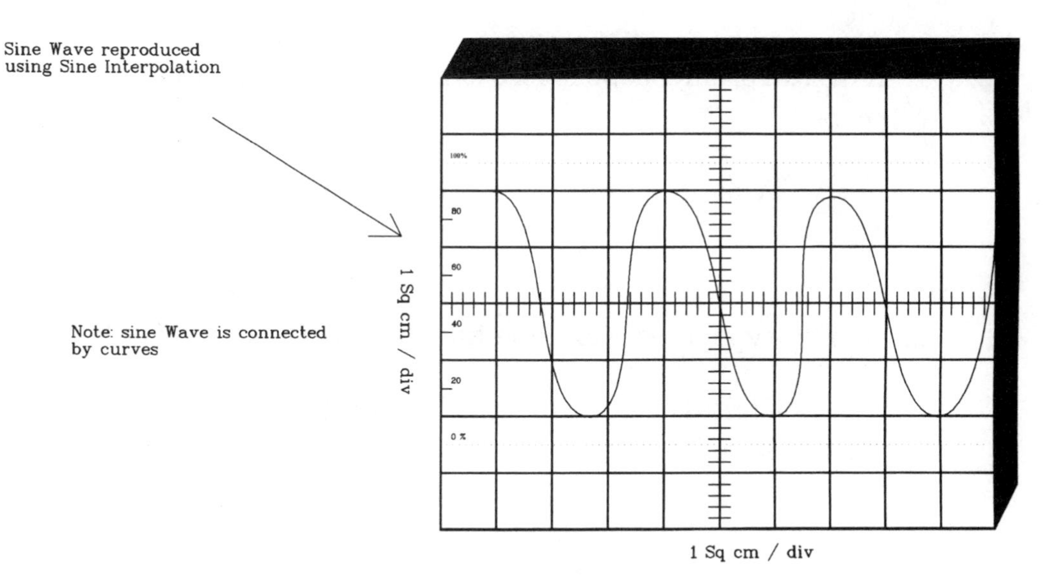

Figure 5-12. Diagram of sine wave interpolation.

Analog oscilloscopes are mostly found in electronic schools and services facilities.

Digital oscilloscopes are used for both analog and digital work.

Interpolation is a technique used to reconstruct a waveform of a fast signal.

Digital oscilloscopes use what are called ADC and DAC devices to process digital data.

Digital oscilloscopes are used mostly for viewing one-shot events or transient signals.

CHAPTER 5 QUIZ

1. A digital oscilloscope is less expensive than an analog oscilloscope?

   A. True　　B. False

2. Which type of oscilloscope has more bells and whistles?

   A. Digital　　B. Analog
   C. Same

3. What type of oscilloscope would you use to measure fast moving signals?

   A. Digital　　B. Analog
   C. Neither

4. Digital oscilloscopes are best used for what application?

   A. One-shot　　B. RF
   C. Low frequency　　D. High frequency

5. Digital oscilloscopes use what type of sampling?

   A. Interpolation　　B. Real time
   C. Real mode　　D. None of the above

6. The digital oscilloscope's ADC is used for?

   A. Analog to 1's and 0's   B. Keep the CPU cool
   C. Convert 1's to 5's   D. None of the above

7. The digital oscilloscope's sample rate is determined by the _______?

   A. Vertical section   B. Trigger
   C. Horizontal system   D. None of the above

8. Most small service facilities have a(n) ______ oscilloscope?

   A. Digital   B. Analog
   C. Neither

9. Which type of oscilloscope costs more money?

   A. Analog   B. Digital
   C. Same

10. Low signals are best seen on a(n) ______ type oscilloscope?

   A. Analog   B. Digital
   C. Either   D. Neither

11. What is common to both oscilloscopes' vertical amplifiers?

    A. Probe B. Power supply
    C. Horizontal system D. None of the above

12. What type of oscilloscope should be used to see a transient response?

    A. Analog B. Digital
    C. Either D. Neither

13. Oscilloscopes use a(n) _____ to test the circuits?

    A. Battery B. Cable
    C. Probe D. Antenna

14. Digital oscilloscopes are less complex than analog type scopes.

    A. True B. False

15. Analog oscilloscopes use memory to store waveforms.

    A. True B. False

NOTES

CHAPTER 6
OSCILLOSCOPE GROUNDING

CONTENTS:

Setting up your test bench is the final stage to making successful electronic measurements and tests. The service bench setup should provide the user with all of the necessary test equipment for making good and accurate measurements, and keeping potentially dangerous voltages from traveling though you causing an electrical shock and damaging sensitive circuits, like ICs. This chapter briefly describes how to properly set up a work station, and specifically how to connect your oscilloscope and other test equipment to ground. This section will show you how to protect yourself from hazardous shocks, and describe one method used for grounding yourself.. Grounding yourself protects circuits from high static charges that can ruin expensive ICs or other types of static sensitive devices.

GROUNDING EQUIPMENT

Grounding electronic equipment is an important step when setting up your test bench, because it can save your life. Without grounding, high voltages and currents can travel through your body and cause death. Properly grounding your oscilloscope and is necessary for good safety. See **Figure 6-1**.

Most modern oscilloscopes and electronic test equipment are well insulated. Although the case and knobs may appear safe, high voltages and currents can travel though these materials causing burns in some situations. Only by grounding the equipment, either by connecting a wire of proper size from the equipment's metal chassis to a good earth ground, or using the equipment's three-pronged power cord plugged into an outlet grounded to earth, are you protected from these potentially dangerous voltages and currents. When equipment is grounded in this fashion it is called *single-ended.*

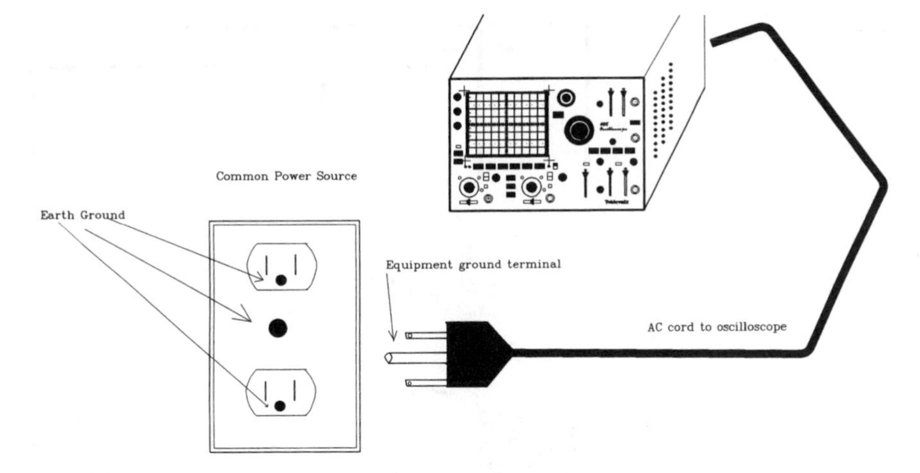

Figure 6-1. Diagram of receptacle and equipment ground via AC plug.

Note: Not all equipment require grounding. Battery operated oscilloscopes and other non-AC type equipment that use high voltages internally are well insulated to protect you from being shocked. However, always practice safety first.

When setting up your test bench make sure that all equipment share the same ground. This is a necessary and important step because it insures that the equipment and oscilloscope will make accurate measurements. It will also keep the ground leads from shorting out other equipment and circuits under test. See **Figure 6-2**.

The oscilloscope must use the same ground as the circuit. If the circuit is floating, or not at the same ground level, damage could result to the circuit and possibly to the test equipment as well, as seen in **Figure 6-3**.

The circuit grounding is of vital importance, because it sets the voltage references for all the measured voltages on the circuit, and serves as the platform reference for all other test equipment in the chain.

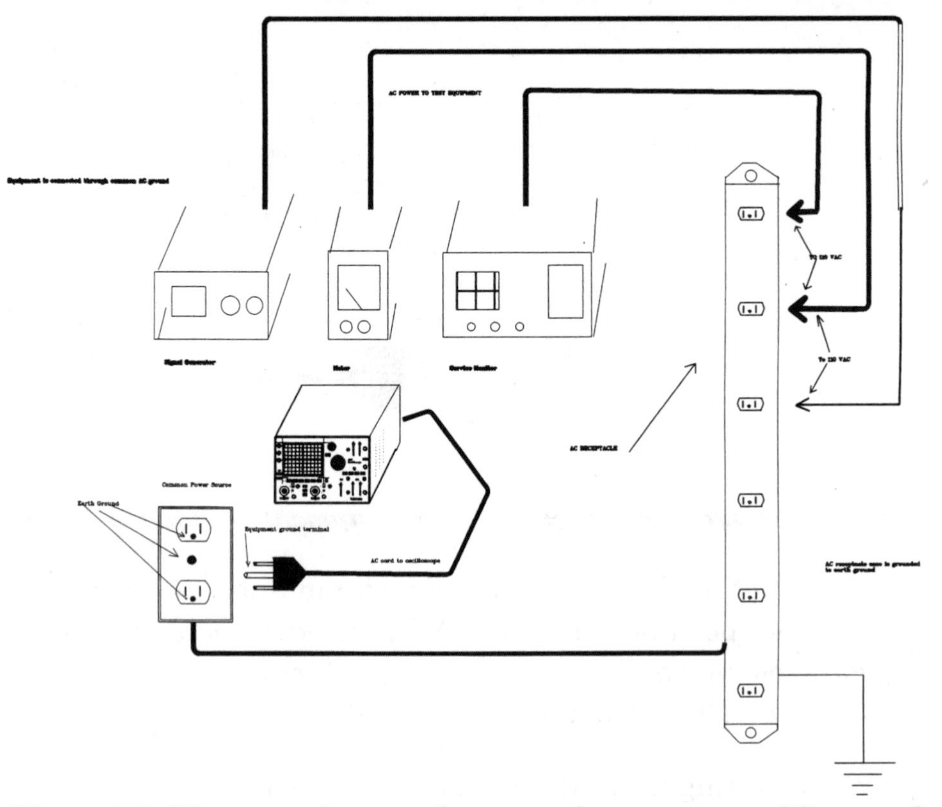

Figure 6-2. Diagram of test equipment using a common AC ground.

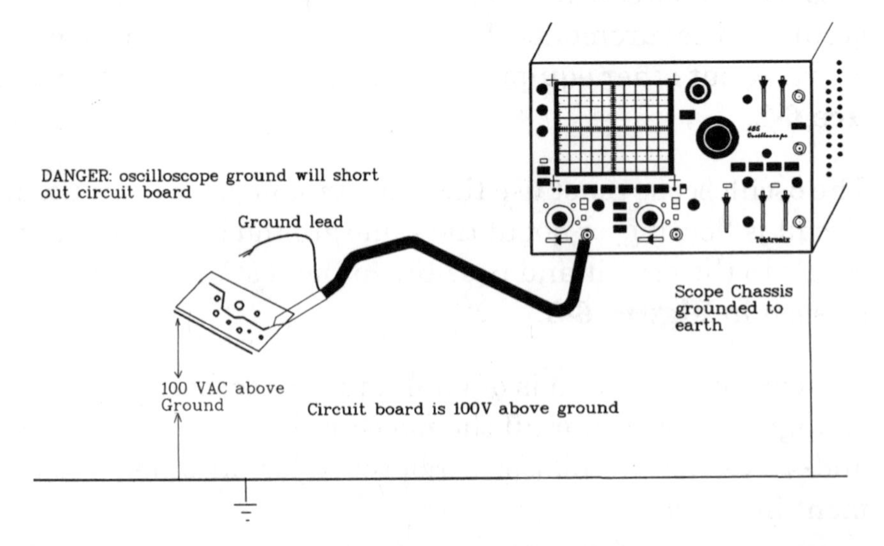

Figure 6-3. Diagram of floating circuit with earth grounded scope.

GROUNDING YOURSELF

Grounding yourself is important when working with any device or components that are static-sensitive. Many manufacturers who supply these devices usually pack them in special packaging materials that are labeled, CAUTION STATIC-SENSITIVE. These devices can easily be ruined by static electricity. The conditions of the environment or the type of clothes worn can build up a dangerous static charge on your body, which could ruin an expensive IC if you came in contact with it by touching the leads.

When using your oscilloscope to repair circuits that have ICs or other static-sensitive devices, always make sure you are grounded via a ground strap. The ground strap helps to reduce the static build up on your body by sending static charges to earth ground. **Figure 6-4** shows a typical wrist-type ground strap worn by engineers and technicians when working with static devices.

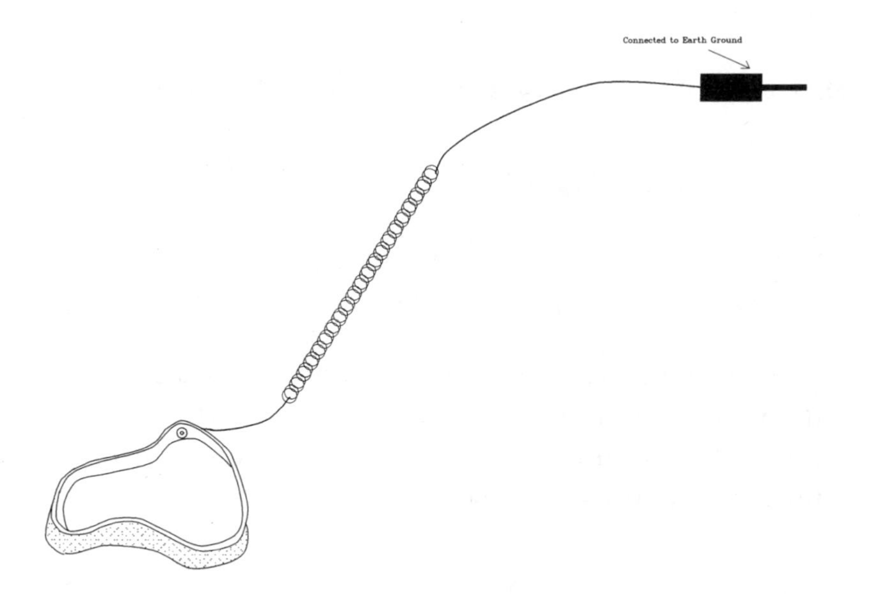

Figure 6-4. Typical wrist type grounding strap.

Ground straps are usually worn on the wrist closest to the hand that comes in contact with the device. The other end of the ground strap must be connected to earth ground in order to breed the static charges to ground. See **Figure 6-5**.

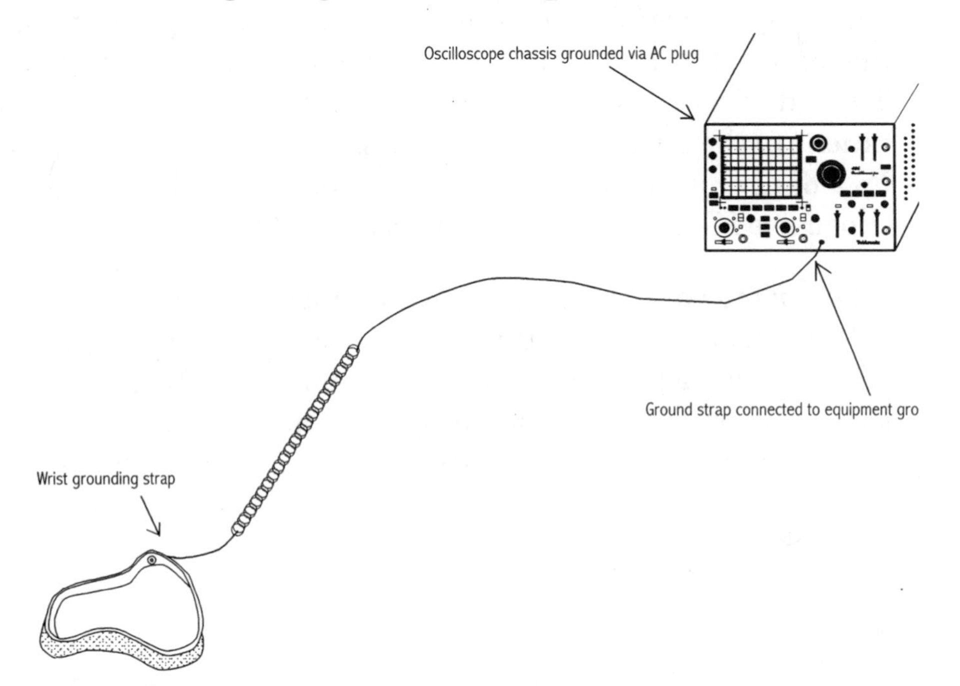

Figure 6-5. Ground strap connected to oscilloscope's ground lug.

FLOATING OSCILLOSCOPES

Caution: Conventional oscilloscopes in single-ended configuration can only make single-ended measurements.

A serious electrical safety hazard may exist at your test bench if you float your oscilloscope. *Floating* is the intentional severing of the oscilloscopes's ground reference in order to make measurements not of the same ground level.

Many technicians make this simple but deadly modification to their oscilloscope. In doing so, the oscilloscope user would be able to measure with respect to power system and television voltages without an isolation transformer, which is normally used for measuring equipment with HOT chassis. The floating technique solves the reference voltage problem but gives rise to a serious electroshock hazard. The oscilloscope chassis is now HOT! This is a very dangerous operating mode and is strongly discouraged by all oscilloscope manufacturers.

The reason manufacturers are so strongly opposed to this practice is clear. The oscilloscope's chassis and the body of the person operating a floating oscilloscope are elevated to the circuit reference voltage level, often several hundred volts above earth ground. In this hazardous situation, a slip or an inadvertent touch of another surface can be fatal. Anyone can have a forgetful moment. With a HOT oscilloscope the price payed is unacceptably high. Also family members can touch the oscilloscope, unknowingly exposing themselves to serious electroshock hazard.

Note: The level of risk associated with floating an oscilloscope is not warranted.

FLOATING MEASUREMENTS & GROUND LOOPS

When an oscilloscope is used for measurements and its test probe touches a point in a circuit, a waveform usually appears on the oscilloscope screen, even if the ground lead is not connected. In this situation, the reference for measurements is conducted through the safety ground of the scope chassis to the electrical ground in the circuit.

In contrast to oscilloscopes, digital voltmeters also use two probes to measure potentials between two points in a circuit. Because voltmeters are isolated, these two points can be anywhere in the circuit. Historically, this has not always been the case. Before the introduction of high-impedance meters, low-impedances Volt-Ohmmeters (VOMs) were used to measure floating circuits. However, because they had very low impedances it made some circuit measurements difficult, and at other times impossible due to circuit loading. The vacuum tube voltmeter (VTVM) solved most problems of the VOM, but it could not make floating measurements because the VTVM chassis was at ground. Today, most digital voltmeters are isolated from ground and present a high impedance to circuits, and can easily make floating measurements.

Like the VTVM, most bench-type oscilloscopes can only measure voltages that are referenced to earth ground, which is connected to the oscilloscope and other equipment ground points. This ground point connection is called *single-point ground* and is shown in **Figure 6-6**.

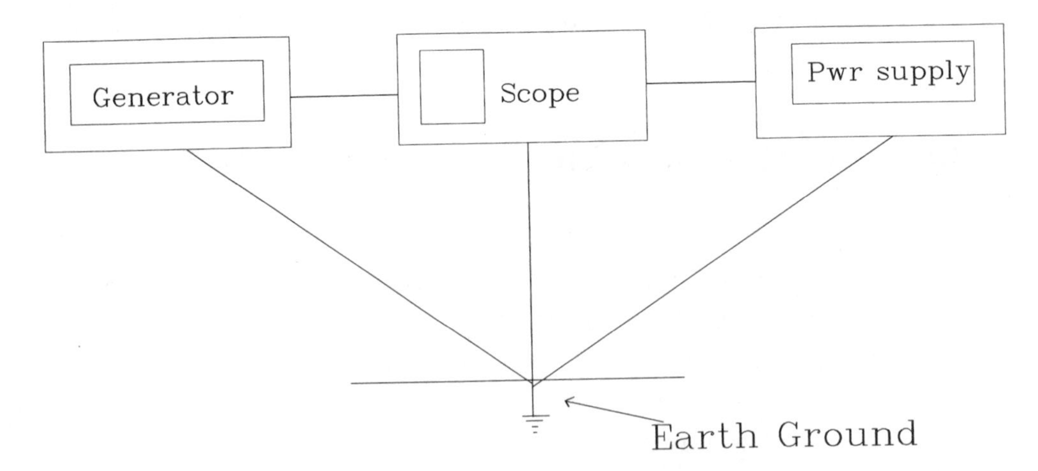

Figure 6-6. Single point ground connection.

The single-point ground is a single physical point in a circuit. By connecting all grounds to a common point, no interference will be produced in the equipment because the configuration does not result in potential differences across the equipment. However, at high frequencies interference can be a problem unless care is taken to reduce capacitive coupling, which will be discussed a little later.

When oscilloscopes and other test equipment are referenced to ground, and measurements must use one probe connected to ground, these types of measurements are called *single-ended measurements* and for obvious reasons.

The probe's ground provides the reference path for the measurements. Unfortunately, there are times when this limitation lowers the integrity of the measurement, or makes measurements impossible using conventional measuring techniques.

When measuring the voltage between two test points, neither of which is grounded, conventional oscilloscope measuring techniques cannot be used because of the potential danger of damaging the test circuit or instrument.

A common example would be the test points TP1 and TP2 in the switching power supply of **Figure 6-7**. This is a simple circuit, which shows some of the complexities that face an oscilloscope user employing conventional measuring techniques.

Single-ended oscilloscopes do not fair well when making balanced signal measurements (between leads without a ground return) like a common telephone line. As we shall see, even some ground-referenced signals cannot be faithfully measured using singled-ended techniques.

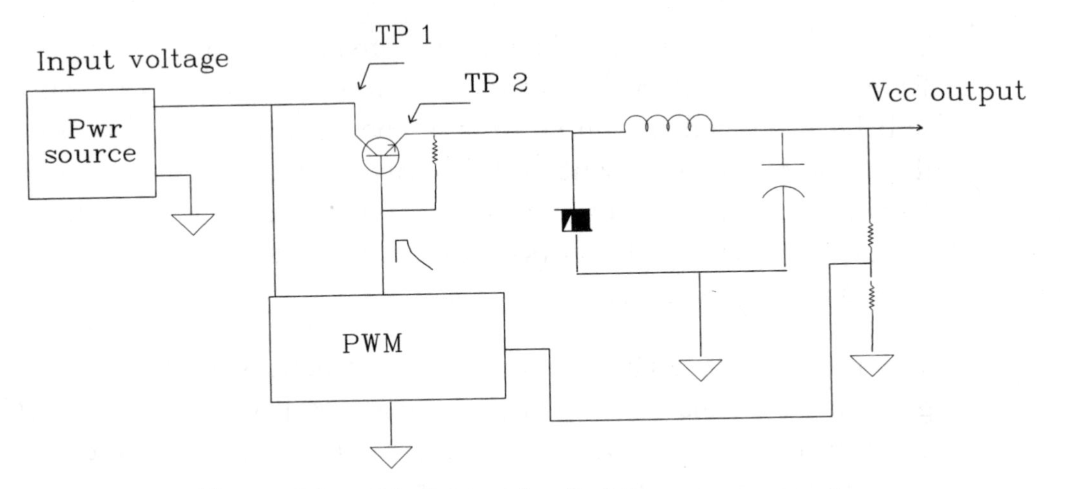

Figure 6-7s. Diagram of switching power supply.

GROUND LOOPS

A ground loop results when two or more separate ground paths are tied together at two or more points. This results in a loop that is highly susceptible to varying magnetic fields. The ground loop acts like the secondary of a transformer which is essentially a shorted turn of low impedance. When a magnetic field excites the loop the current circulating in the loop develops a voltage across any impedance differences within the loop. Thus, at any given instant, various points within a ground loop will not be at the same potential.

Note: A ground loop can be created by any conductor in the vicinity carrying AC current.

When using an oscilloscope to take measurements, connecting the ground lead to the circuit under test can result in a ground loop if the test circuit is grounded to earth ground. See **Figure 6-8**. A voltage potential will be developed in the probe ground path resulting from the circulating current acting on the impedance within the path. If we were to measure the voltage at the oscilloscope BNC connector with an AC meter, we would find the voltage at the BNC connector would not be the same as the circuit ground being measured.

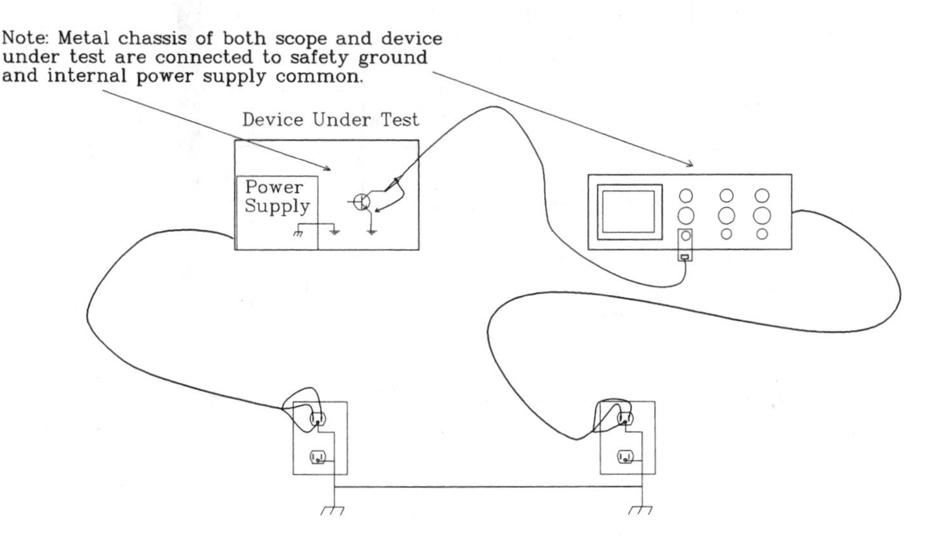

Figure 6-8. Oscilloscope & probe creating ground loop.

This potential difference can range from a few microvolts to as high as hundreds of millivolts. Because the oscilloscope references the measurements from the shell of the input BNC connector, the displayed waveform may not represent the true signal at the probe input. This error becomes more apparent as the signal amplitude decreases, as in noise measurements and biomedical measurements.

If we removed the ground lead when making measurements, the loop would be broken. This technique works, however, only at low frequencies. At higher frequencies, the probe begins to add *ring* to the signal voltage, as seen in **Figure 6-9**. The ringing is caused by self-resonating conditions within the probe via the probe's tip capacitance and shield inductance. See **Figure 6-10**. This is why you should always use the shortest ground leads possible.

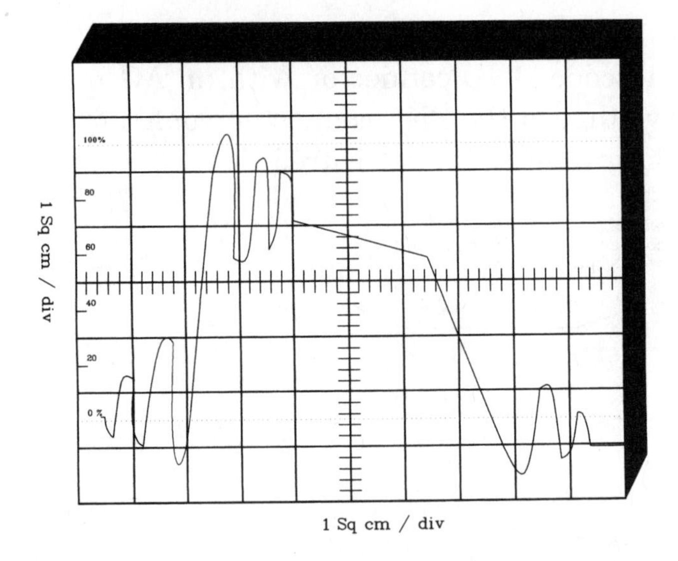

Figure 6-9. Square wave distortion caused by ringing.

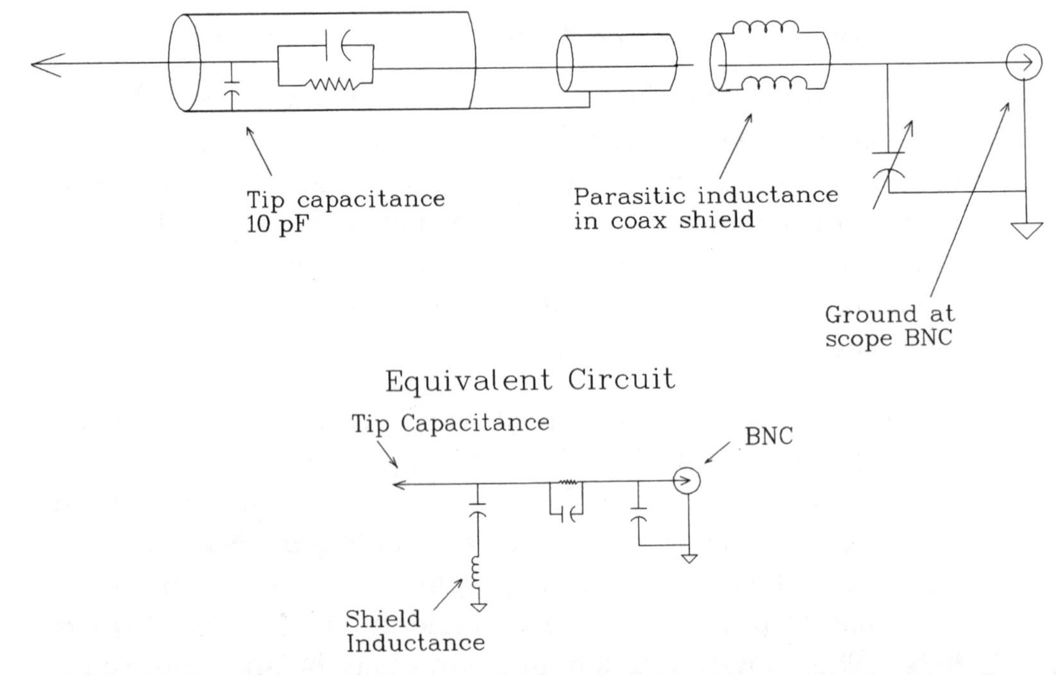

Figure 6-10. Series resonant tank circuit formed by probe tip capacitance and ground inductance.

CATCH 22

How do we deal with the above dilemma: either we create ground loops when we make our measurements and add error into our signal, or we remove the probe ground lead and add ringing to the waveform.

One measurement technique often tried to break ground loops is to float the circuit being measured. Floating either the oscilloscope or the device under test (DUT) allows the use of short ground lead to minimize ring without creating a ground loop. The practice is inherently dangerous, as it defeats the protection from electrical shock in the event of a short in the primary wiring.

Note: Some battery powered oscilloscopes are well suited for floating measurements because they incorporate insulation, which allows for safe floating measurements.

Operator safety can be restored by placing a suitable ground-fault circuit interrupter (GFCI) in the power cord of the oscilloscope (or DUT) with the severed ground. See **Figure 6-11**. However, be aware that without a low-impedance ground connection, radiated and conducted emissions from the oscilloscope may exceed government standards as well as interfere with the measurement itself. At higher frequencies, severing the ground may not break the ground loop as the floating circuit is actually coupled to earth ground through stray capacitance. See **Figure 6-12**.

Note: Floating a digital oscilloscope may cause damage to its electronics.

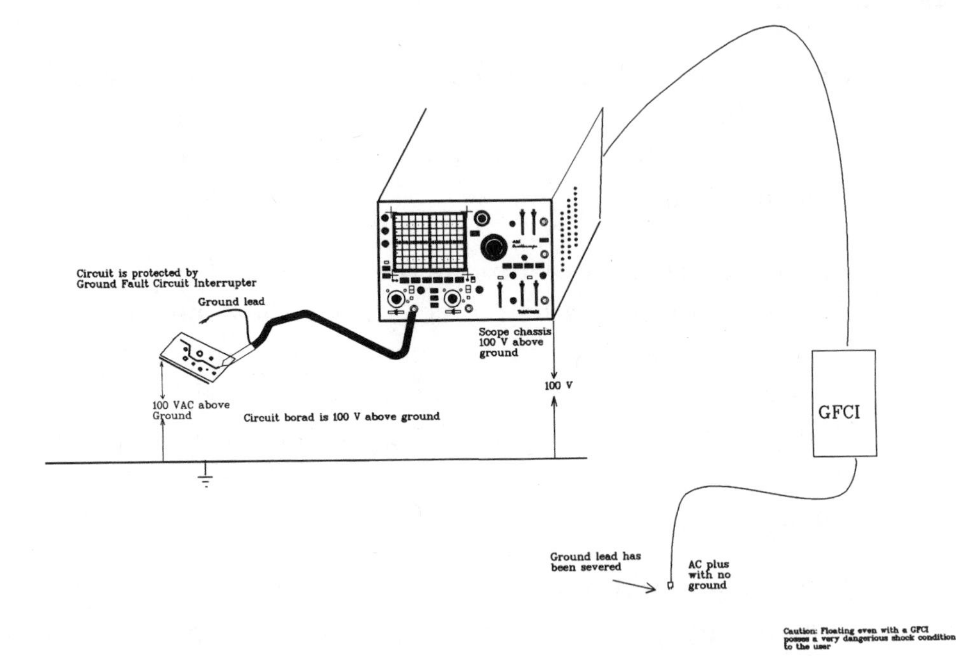

Figure 6-11. Oscilloscope connected to GFCI for ground protection.

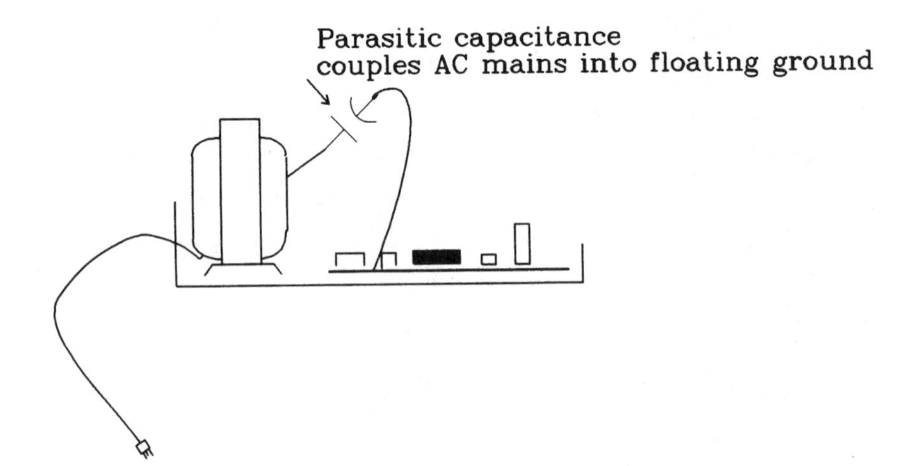

Figure 6-12. Diagram of circuit coupling AC mains into floating ground.

GROUND SYNDROME

Even when the measurement system doesn't introduce ground loops, the "ground is not ground" syndrome may exist within the devices being measured, as seen in **Figure 6-13**. Large static currents and high frequency currents act on the resistive and inductive components of the device ground path to produce voltage gradients. In this situation, the "ground" potential referenced at one point in the circuit will be different than referencing another point.

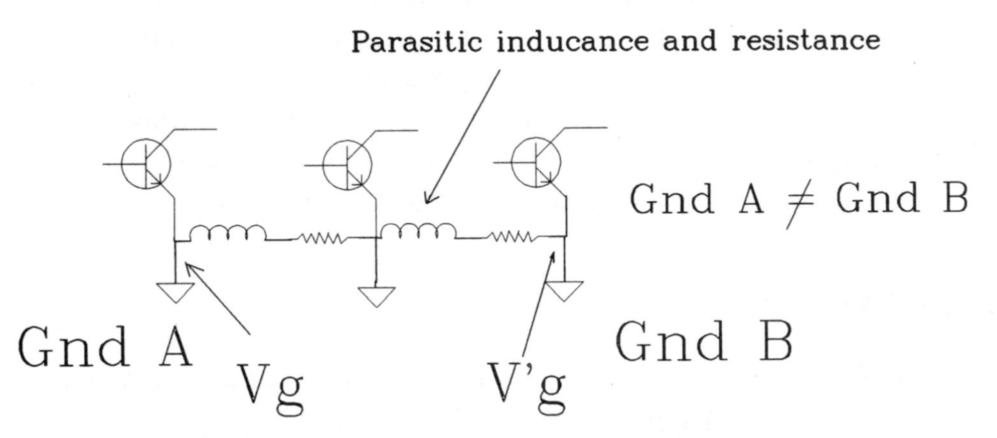

Figure 6-13. Component grounding syndrome effects.

For example, ground at the input of the high-gain amplifier in a system differs from the "ground" potential at the power supply by several millivolts. To accurately measure the input signal seen by the amplifier, the probe must reference the ground at the amplifier input.

These effects have challenged engineers and designers of sensitive analog systems for years. The same effect is seen in fast digital systems. The small inductance within the ground distribution system can create a potential across it, resulting in "ground bounce". Troubleshooting circuits affected by ground-voltage gradients is difficult because of the inability to really look at the signal "seen" at the individual component. No doubt this too will be one of your greatest challenges. Connecting the oscilloscope probe ground lead to a "ground" point of the device results in the uncertainty of what effect the new path adds to the grounds gradient. A sure clue that a change is occurring is seen when the problem in the circuit either gets better (or worse) when the probe ground is connected. What we really need is a method to make an oscilloscope measurement of the actual signal at the input of the suspect device.

The solution to two-point measurements can easily be accomplished with the help of differential probes, and/or isolators. When these devices are used ground loops are eliminated, and no circuit degradations or shock hazards to the user exist. In Chapter 9 we will discuss measuring techniques using differential probes and different methodologies.

SUMMARY

Grounding your test equipment protects you from serious electroshock or death.

A HOT chassis can be dangerous.

Ground straps should always be worn when testing or replacing static sensitive devices.

All equipment grounded must have the same ground reference.

The oscilloscope must use the same circuit ground.

Floating an oscilloscope is strongly discouraged by all oscilloscope manufacturers.

Voltage is always a two-point measurement.

Because a circuit is grounded does not mean it is at ground.

Floating voltages can be measured with a conventional oscilloscope.

Most conventional test equipment cannot be used to measure floating voltages.

Floating voltages can be very dangerous.

Connecting the ground lead of an oscilloscope to ground of the circuit can create a ground loop.

Measuring a floating voltage with a single-ended device may cause equipment damage.

Removing the ground lead on an oscilloscope can create ringing in the waveform.

CHAPTER 6 QUIZ

1. Grounding your oscilloscope is not important for measurements.

   A. True B. False

2. Floating is warranted in some applications of testing?

   A. True B. False

3. The oscilloscope must share the same ground as the test circuit.

   A. True B. False

4. All equipment should be at the same ground level.

   A. True B. False

5. Damage can result if you test a HOT chassis with an oscilloscope probe's grounded lead.

   A. True B. False

6. When replacing static sensitive devices you should always wear _______.

   A. An ankle strap B. A wrist strap
   C. Rubber shoes D. Rubber gloves

7. Portable oscilloscopes and non-AC operated equipment are safe against electrical shock?

   A. True B. False

8. Most bench type oscilloscopes cannot make ______ measurements.

   A. Current B. Floating
   C. Voltage D. Accurate

9. Ground loops can cause _______ to circuit measurements.

   A. Noise B. Hum
   C. Interference D. Damage

10. Disconnecting the ground lead on an oscilloscope can cause ________.

   A. Hum B. Ringing
   C. White noise D. Feedback

11. Floating voltages can be measured by a ________.

   A. Digital voltmeter B. Current meter
   C. Oscilloscope D. Tachometer

12. A conventional oscilloscope can cause damage to a circuit if precautions are not taken.

   A. True B. False

13. Floating as in referring to an oscilloscope means _______.

    A. Breaking the AC ground
    B. Suspending the oscilloscope
    C. Putting the probe in water
    D. Disconnecting the power supply

14. A ground loop is always safe for measurements.

    A. True   B. False

15. Accurate two-point measurements cannot be taken with an oscilloscope.

    A. True   B. False

NOTES

CHAPTER 7
OSCILLOSCOPE PROBES

CONTENTS:

According to the *World Book Dictionary* the definition of *Probe* is: search into; examine thoroughly and investigate. Popular technical literature describes a probe as any conductor that makes a connection between the circuit under test and the measuring instrument. The conductor can be bare wire, meter leads or even coaxial cable used in satellite communications. In this chapter we will investigate the many different types of probes, compensation of probes and their definitions as they apply to oscilloscope use.

ELEMENTARY PROBES

An oscilloscope is only as good as its probes. If we needed to make a probe in an emergency using just two bare wires connected to our test circuit and oscilloscope, what would we be able to tell about that circuit's electrical behavior? If we then switched those wires and used a coax cable or meter leads in its place, would the waveform change?

For many years oscilloscope designers and manufacturers asked the same questions. The early analog oscilloscopes of the 60's used two bare wires usually colored red and black. The red lead had an alligator clip attached to it, and the black lead was the ground lead. It was soon found by engineers that this type of probe caused many circuit problems. Oscilloscope users were unable to measure high-impedance circuits with their probes without changing the test circuit's electrical characteristics. It was learned that bare wires had high capacitance and inductance, and could even cause a short circuit at certain frequencies. This unwanted change in circuit behavior is called *circuit loading* and these type of probes would be called *elementary probes*.

The investigation continued, trying other types of wire, only to learn that many types of unshielded wire were often susceptible to stray signal pickup. It wasn't uncommon to pick up AM radio broadcasts during measurements.

Soon after these AM broadcasts, many manufacturers used coaxial cable as a means of connecting the test circuit to the oscilloscope, but soon found that coax severely loaded down the circuit, and created resonant problems when measuring signals at certain frequencies. Standing waves were created on the line and the coax cable effectively became a filter giving erroneous measurements and results.

TODAY'S TEST PROBES

Today's test probes are more sophisticated. Designers and engineers have solved many of the problems of yesterday's probes. Using today's technology, newer probes have greatly minimized many of the negative effects associated with the older types of test probes. High speed digital signals and high frequencies have required constant efforts on probe design, producing more than fifteen basic oscilloscope probe types and definitions.

PROBE TYPES

Probes come in a variety of shapes and colors. Each oscilloscope manufacturer usually packs a set of general purpose probes with your oscilloscope so you can begin using it right away. The type of probe that you use is usually indicative of the test application. The most popular is the general purpose X10 probe. See **Figure 7-1**. The X10 probes are used for general oscilloscope use, and provide a convenient way to get you started. They are usually passive, so they do not require power for their operation.

Probes can be classified by the type of signals they will be used to measure. For example:

Voltage
Current
Logic
Temperature
Light

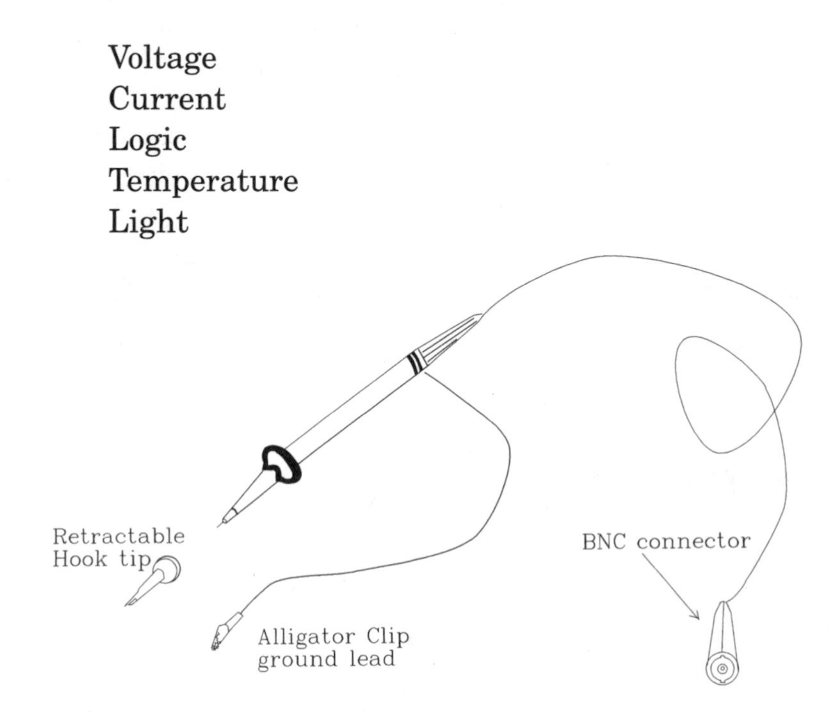

Figure 7-1. Diagram of general probe.

Probes are either passive or active; the type will depend on your specific application. The following list shows the different types of probes used in oscilloscope measurements. See **Figure 7-2**.

General Purpose Voltage Probes
High Impedance Probes
Low Impedance Probes (50 ohm)
High Voltage Probes
Current Probes
Temperature Probes
Logic Probes
Differential Probes

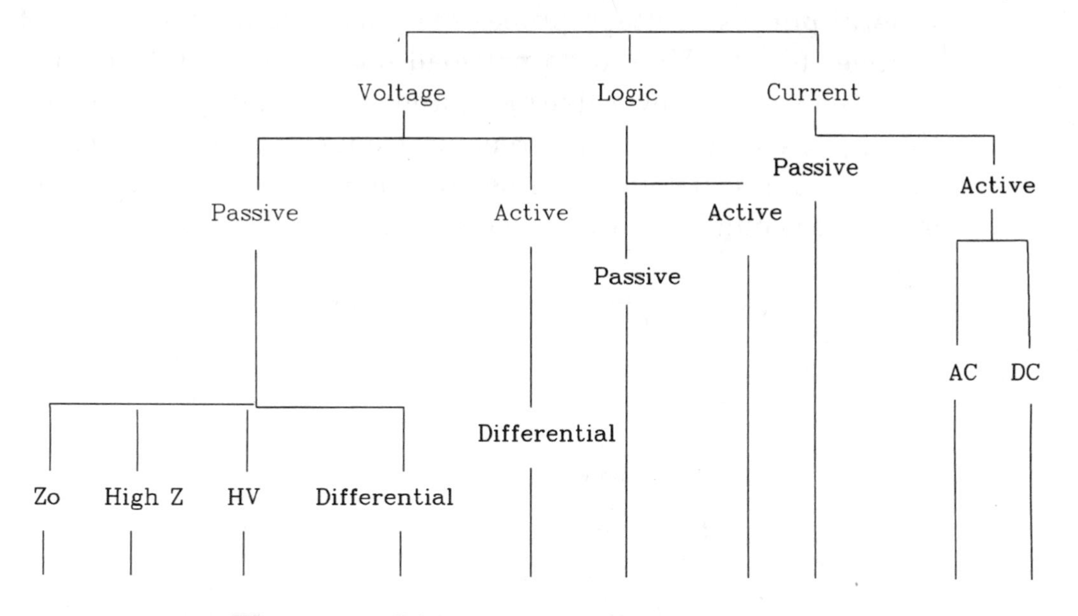

Figure 7-2. Diagram of basic probe families.

GENERAL PURPOSE PROBES

Most oscilloscope users use the general purpose X10 attenuating probe for measurements. The number before the variable X denotes the probe as an attenuation type probe, while a probe with the variable after the X would be an amplifying probe. The attentuation probe reduces the signal by the noted factor; for example, if testing a signal measuring 10 sq divisions on the display, using an X10 probe would reduce this measurement to 1 division. This is usually the type of probe you will get with your oscilloscope. The X10 general purpose probe can measure high frequencies and maintain the integrity of your test circuit by minimally loading the circuit.

General purpose probes can have attenuation factors as high as 1000. See **Figure 3**. General purpose probes are inexpensive and are easy to use with only minimum balancing. Most general purpose voltage probes can measure a wide range of voltages (+/-400 V) and do not load down the circuit below 5 MHz or lower. However, like most probes their capacitance across the input changes with increasing frequency. As the capacitance changes the output impedance loads the test circuit creating a profound effect on the test circuit performance.

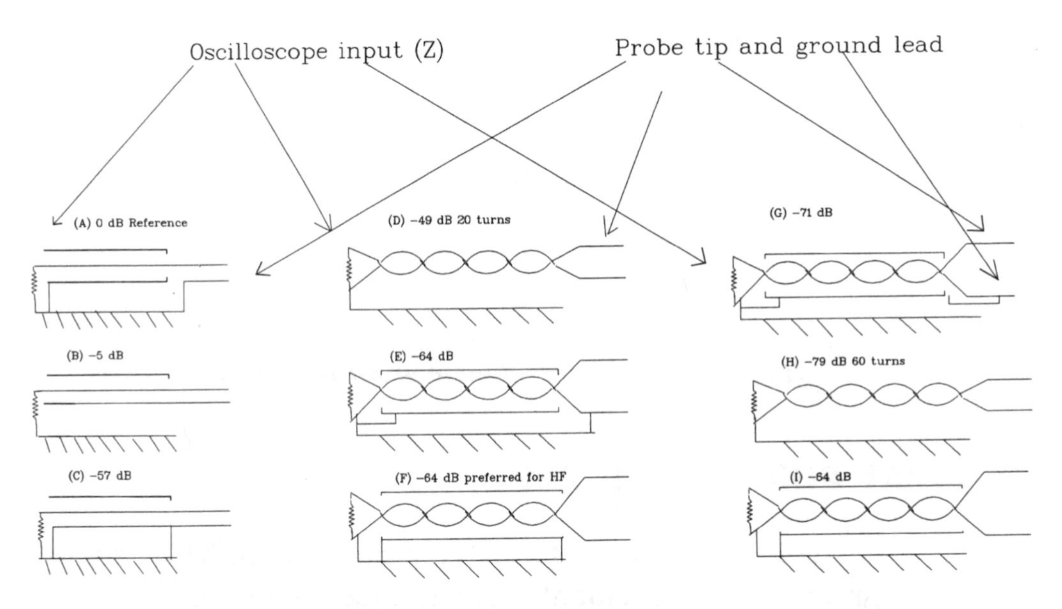

Figure 7-3. Diagram of elementary probe types and attentuation factors.

HIGH IMPEDANCE ACTIVE PROBES

High impedance active probes offer the user of general purpose probes many advantages not found when using the general purpose probe. Although active probes require power for their operation and can be very expensive, the input and output impedance is maintained throughout the measurement, with no signal loss, and high input impedance (typically >100 Kohms) is maintained. Active probes incorporate a semiconductor or FET as an impedance converter. The input of the FET is high while the output is low (usually 50 ohms), which makes this type of probe a good 50-ohm line driver, making remote measurements possible.

LOW IMPEDANCE 50-OHM PROBES

Some oscilloscopes have a 50-ohm input. Low impedance 50-ohm probes work best for very high frequencies because their output capacitance is usually the lowest of all probe types, and provide the most consistent loading of any probe family.

HIGH VOLTAGE PROBES

High voltage probes are used for making high voltage measurements. The attenuating factors can be as high as 100 or 1000, depending on the voltage you are measuring and the circuits loading characteristic.

CURRENT PROBES

The current probe can be viewed as the black sheep of the family of probes; it looks like no other testing probe. The probe does not have to make a direct connection to the circuit voltage or current in order to make a measurement. The current probe couples energy directly from the surrounding fields. Current probes come in two types, AC and DC, and can measure AC signals from a few cycles to 2 GHz. The DC versions can measure from DC to as high as 50 MHz. The current probe, because it does not directly interfere with the circuit voltages or currents, creates the least circuit loading of any other probe.

TEMPERATURE PROBES

Temperature probes can be utilized in a number of different circumstances; such as when you wish to monitor ambient temperature, record changes of temperature during experiments, or note breakdown temperatures of circuits. Most probes utilize either thermistors, passive semiconductors which change resistance with temperature changes, or thermocouples, which employ the Seebeck Effect, producing an EMF between two dissimilar metals.

LOGIC PROBES

Logic probes provide a quick and inexpensive way to troubleshoot digital equipment. They are compact, self-contained test instruments, usually with two or three LEDs to indicate high, low, or alternating digital states. Although they don't show actual voltages, logic probes enable the technician to check pins on crowded circuit boards for their proper logic states, without having to use more expensive and bulky instruments. If a circuit is toggling between high/low states too quickly to be discerned with the LEDs, though, an oscilloscope will be needed to properly view the signal.

DIFFERENTIAL PROBES

Differential probes come in handy when you are making noise or other measurements specifically requiring common mode rejection (CMR), i.e., noise measurements. Differential probes are designed in two types, active and passive. Active type probes have a differential amplifier built in the tip of the probe and require a power source for their operation. Passive type probes require a differential amplifier incorporated in the input channels of the oscilloscope.

PROBE SELECTION

Now that we know a little about probes, we can choose from the many different types of probes on the market. When using your oscilloscope you must decide which probe is best for the job. When making your selection it helps to understand each probe's characteristics, and how each probe affects the circuit behavior, while keeping in mind that proper probe selection will enhance the oscilloscope waveform, and the wrong probe will reduce your equipment's viewing capabilities of that same waveform. While the main consideration for an appropriate probe is its loading and signal transfer characteristics, the probe's physical parameters also play an important part in the selection process. Cable length, probe size and adapters should also be taken into consideration during your probe selection.

BANDWIDTH AND RISE TIME

Probes, like other RF devices, have bandwidth. Your probe's bandwidth should be the same as your oscilloscope's bandwidth. The bandwidth of your probe is an indication of what you can expect your probe to accurately measure in terms of frequency with only a -3dB loss in signal amplitude, or 0.707 of the maximum frequency signal level. In most voltage probes, the bandwidth/rise time product is close to 0.35 and is verified by a pulse rise time to ensure minimum aberrations. Most probe manufactures use a 50-ohm system to define these parameters accurately when terminated in 50 ohms or lower.

PROBE MATCHING

In order to transfer maximum signal energy from the test circuit with the least amount of signal loss, the probes must be matched. When correctly matched in an equal impedance the probe will have a constant *attenuation ratio* throughout its bandwidth, and standing waves on the coaxial cable will be at minimum, reducing circuit loading. When selecting the proper probe always make sure the probe matches the oscilloscope input impedance and capacitance. For example, if the oscilloscope you are using has an input impedance of 1 Megohm and 4 pF of capacitance, make sure the probe you choose also has a 1 Megohm input, and has 4 pF of capacitance. If the oscilloscope has a 50-ohm input and 10 pF make sure the probe you plan to use has the same 50-ohm and 10-pF markings. See **Figure 7-4**. The probes must be the same impedance as the oscilloscope for maximum efficiency.

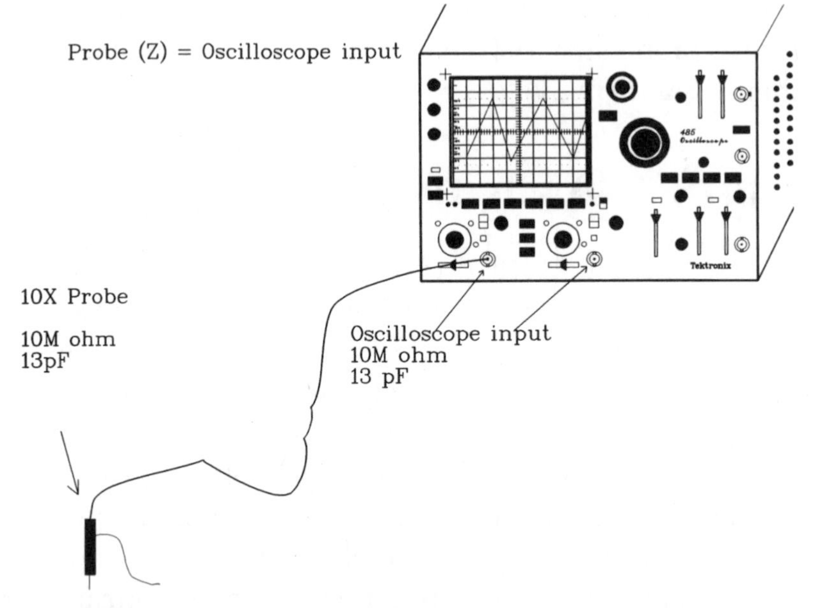

Figure 7-4. Probe matches oscilloscope input (Z).

MAXIMUM VOLTAGE

Oscilloscope manufacturers rate their probes as (DC + peak AC) maximums, but when probes aren't matched these maximum voltages become greatly reduced because of reflections on the cable, causing uneven voltages and currents resulting in lower voltage breakdowns. When a probe is matched, the voltages and probe bandwidth are usually at their maximum. However, as the probe impedance begins to change due to changing frequencies voltage derating occurs. See **Figure 7-5**.

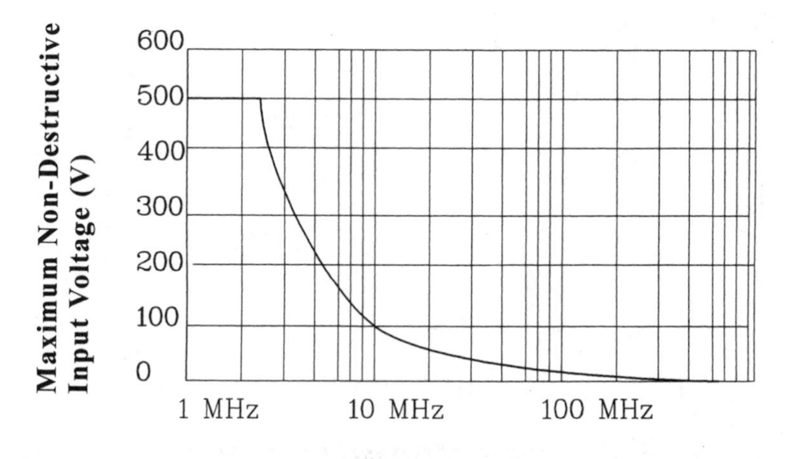

Figure 7-5. Probe voltage derating curve.

IMPEDANCE CHANGES

Probes are not perfect, nor are the components that make the probe. If capacitors did not have reactance, probes would not change impedance with frequency. However, because we are in a real world and capacitors do have reactance, we must somehow deal with the fact that probes change their input impedance. The changing reactance with frequency affects the probe input impedance.

Since input capacitance plays such an important role you should carefully consider this value when selecting your probe. Looking at a graph and probe schematic will help to understand how the magnitude of the impedance changes with changing frequencies. **Figure 7-6** is a schematic of a typical high-impedance X10 probe. **Figure 7-7** shows the relation of impedance changes to frequency.

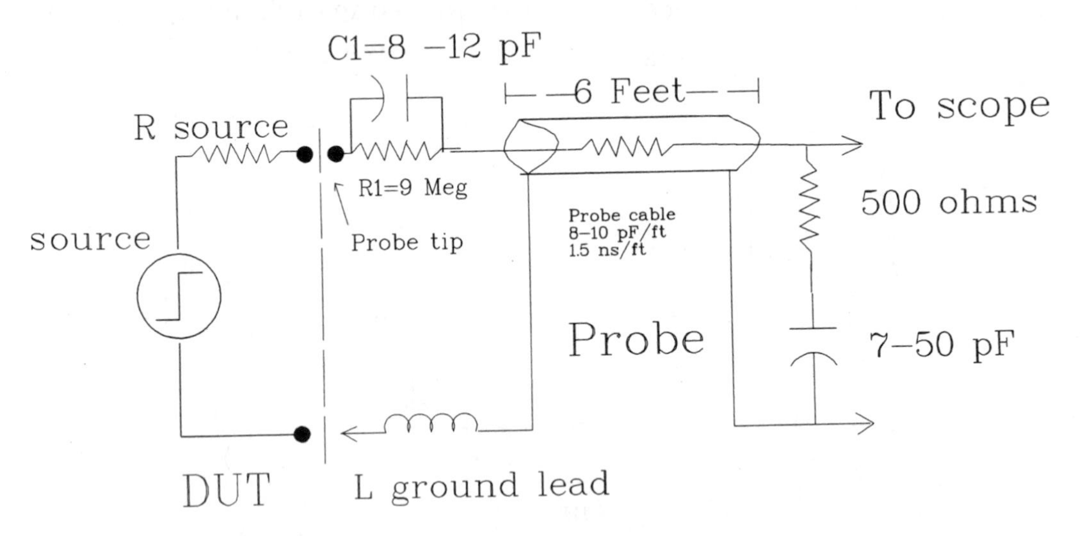

Figure 7-6. Schematic of typical high (Z) passive probe.

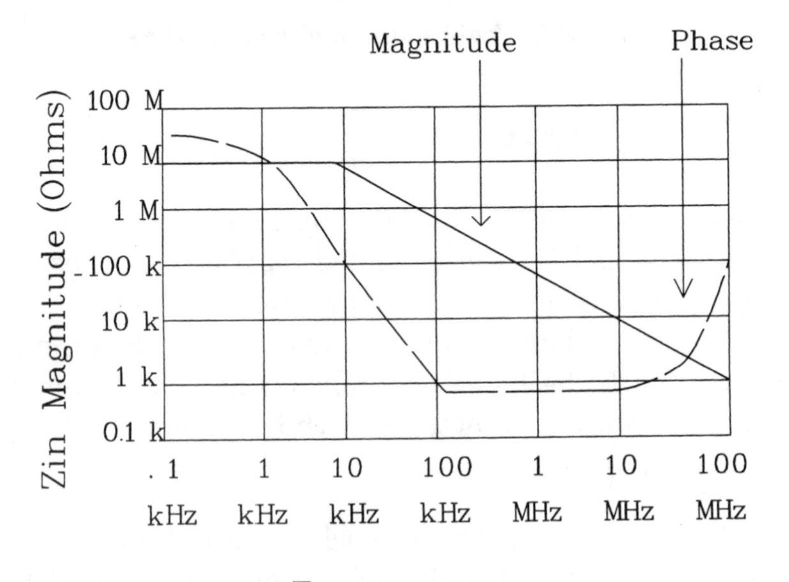

Figure 7-7. Graph of changing input (Z) with frequency.

COMPENSATION

Probes without compensation can have a devastating effect on waveforms. All users of oscilloscopes should get into the habit of balancing (compensating) their probes any time they replace a probe or set up an oscilloscope. The following diagrams in **Figure 7-8** show how compensation affects what you see on your oscilloscope. When your probe is correctly compensated your measurements will be at optimum.

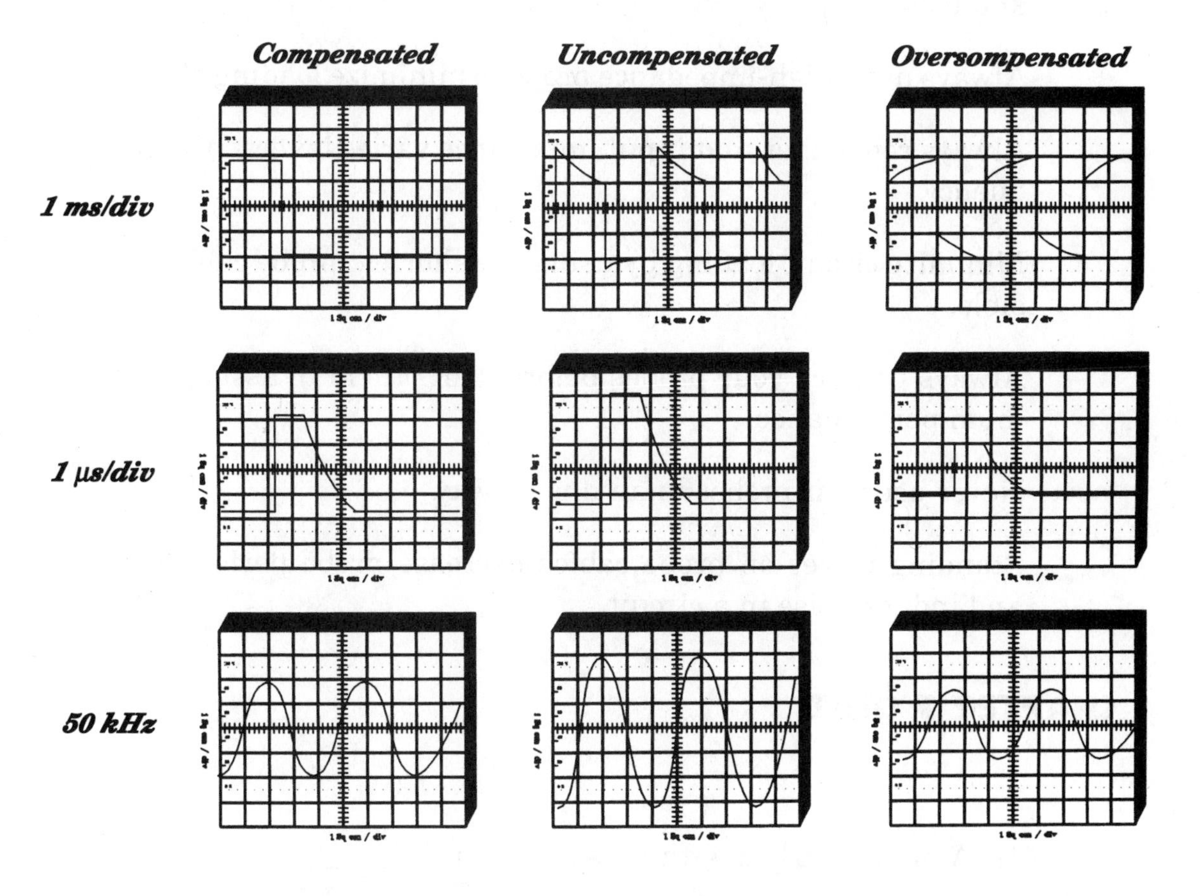

Figure 7-8. Compensation effects on input signal.

SUMMARY

Always match the probe's input resistance and input capacitance to the oscilloscope.

Always match the probe's bandwidth to that of the oscilloscope bandwidth.

Probes should be chosen according to the circuit being measured and tested.

Always use a high-impedance probe to minimize loading effects.

Always remember that input impedance varies inversely to frequency.

Minimize circuit loading by using the shortest probe cable possible.

Always balance your probes before using them to assure optimum performance.

Never use your probes in an unsafe way.

Standing waves on probe cables decrease probe performance and induce noise in a circuit.

CHAPTER 7 QUIZ

1. A probe marked with an X10 is a(n) _________?

   A. Attenuation probe   B. Amplifying probe
   C. Temperature probe   D. Logic probe

2. A probe marked with the 10X is a(n) ________?

   A. Temperature probe
   B. Attenuation probe
   C. Logic probe
   D. Amplifying probe

3. Probe impedance changes with ________?

   A. Temperature
   B. Frequency
   C. Time
   D. Amplitude

4. A voltage probe used for measuring high voltages is usually active?

   A. True
   B. False

5. Passive type probes are inferior to active type probes?

   A. True
   B. False

6. A probe used in general measurements is called a ________?

   A. Common probe
   B. Simple probe
   C. Complex probe
   D. General purpose probe

7. What type of probe offers the least amount of circuit loading.

   A. Passive probe
   B. Active probe
   C. Current probe
   D. Logic probe

8. To help reduce circuit loading the user should use a probe with a ______?

   A. Short ground B. Long ground
   C. No ground D. None of the above

9. All capacitors have ________?

   A. Resistance B. Reactance
   C. Current flowing D. None of the above

10. Bare wires can act like antennas when used as test probes?

    A. True B. False

11. Bare wire makes the best oscilloscope probes?

    A. True B. False

12. Oscilloscope probes must be _______ before using them on a test circuit?

    A. Balanced B. Tested
    C. Removed D. None of the above

13. Coaxial is the best conductor for transferring test signals?

    A. True B. False

14. What type of probes are best suited to pick up stray signals?

    A. Test probes
    B. Elementary probes
    C. Voltage probes
    D. Passive probes

15. Circuit loading is a desirable effect?

    A. True
    B. False

NOTES

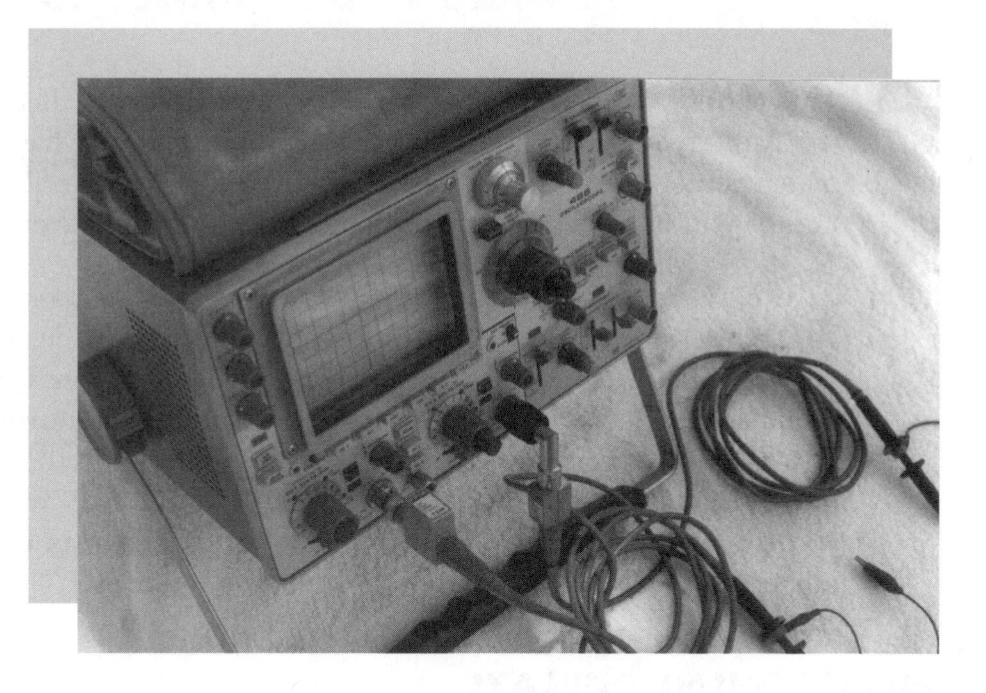

CHAPTER 8
MEASUREMENTS

CONTENTS:

This chapter discusses basic measuring techniques. It teaches the fundamental basics of making good voltage and frequency measurements. Just about every measurement an oscilloscope makes will be based on one of these two basic measurements.

In the previous chapters the illustrated oscilloscope was the Tektronix 485, a manual oscilloscope. No software is used to help set up the viewing of waveforms. Many newer types of oscilloscopes use internal software that make measurements automatically. For the benefit of the new users of oscilloscopes, knowing how to first take measurements manually will help you understand and check automatic measurements of more advanced oscilloscopes models.

MEASUREMENT DISPLAYS

Let's take a closer look at the oscilloscope. In Chapter two, we called the grid markings on the screen *graticules*. We also noted that each vertical and horizontal line which formed a square box, constituted a *major division*. In most oscilloscopes the graticule is usually back-lighted for easier viewing and laid out in an 8 by 10 division pattern as seen in the earlier diagrams. The oscilloscope controls that are labeled (Volts/Div and Sec/Div) always refers to major divisions. Looking closer at the center of the display you will note small tick marks on the horizontal and vertical graticule lines. The tick marks are called *minor divisions*. See **Figure 8-1**.

The display also has percentage markers on the left side of the display. These markers are to help the user make rise time measurements, discussed later in the chapter.

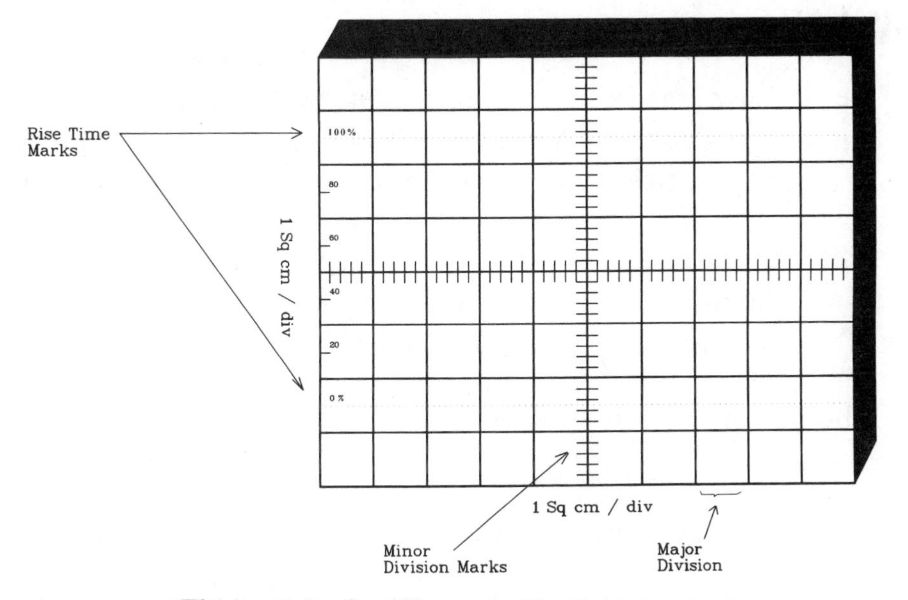

Figure 8-1. Oscilloscope display graticule.

VOLTAGE MEASUREMENTS

Oscilloscopes measure voltages. Voltages are always measured between two points in a circuit. This is true whether using a voltmeter or an oscilloscope. When an oscilloscope probe touches a point in a circuit a waveform is usually seen. Usually one of these points is ground (zero volts) but not always (a ground is not always a ground as will be pointed out in the next chapter). Voltages can be measured in peak to peak values, peak values, and RMS values. You must be careful when making these voltage measurements and specify which voltage values you mean when making measurements.

The oscilloscope does a poor job of measuring current with a voltage probe. The voltage probe is the only device for accurately transferring voltage signals from the circuit to the instrument. Once the voltage measurements are made via the probe, other quantities are just a calculation away.

OHMS LAW

Ohms Law states: a voltage measured between two points is equal to the current times the resistance. By knowing any two of these quantities you can calculate the third quantity.

Voltage = Current x Resistance	or	E = I x R
Current = Voltage ÷ Resistance	or	I = E÷R
Resistance = Voltage ÷ Current	or	R = R÷I

The power law calculations can also be applied to AC and DC circuits.

Power = Voltages time Current	or	P = E x I

Calculations for AC signals are more complicated; however, the point here is that measuring the voltage is the first step towards calculating other quantities. **Figure 8-2** shows the volt-

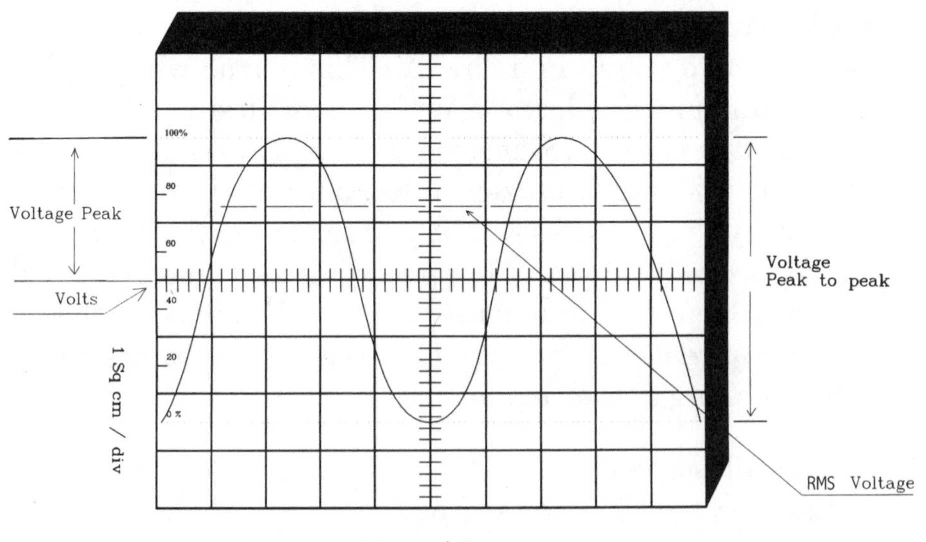

Figure 8-2. Diagram of peak to peak and RMS voltage values.

age of one peak (Pv) and the peak to peak voltages. Use the RMS (root mean square) voltage (Vrms) to calculate the power of an AC signal.

ACCURATE MEASUREMENTS

The greater the area the waveform covers on the screen, the more accurate the measurement will be. Adjusting the signal to cover most of the screen vertically, then taking the measurement along the center vertical graticule line which has the smaller divisions, makes for the best voltage measurements. Always take amplitude measurements at the center of the vertical graticule line. See **Figure 8-3**.

Many newer oscilloscopes have on-screen cursors that let you take waveform measurements automatically on the screen, without having to count graticule marks. Basically, these cursors are two horizontal lines for voltage measurements, and two ver-

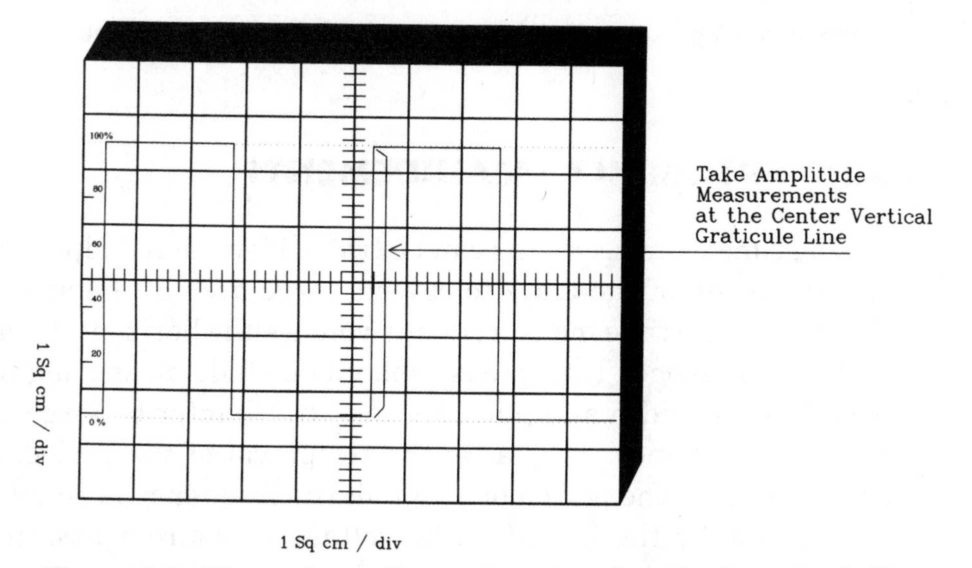

Figure 8-3. Measuring voltage at center of vertical graticule line.

tical lines for time measurements which can be moved around the screen. A readout shows the voltage or time at their positions, and displays the results on the side of the CRT display. See **Figure 8-4**.

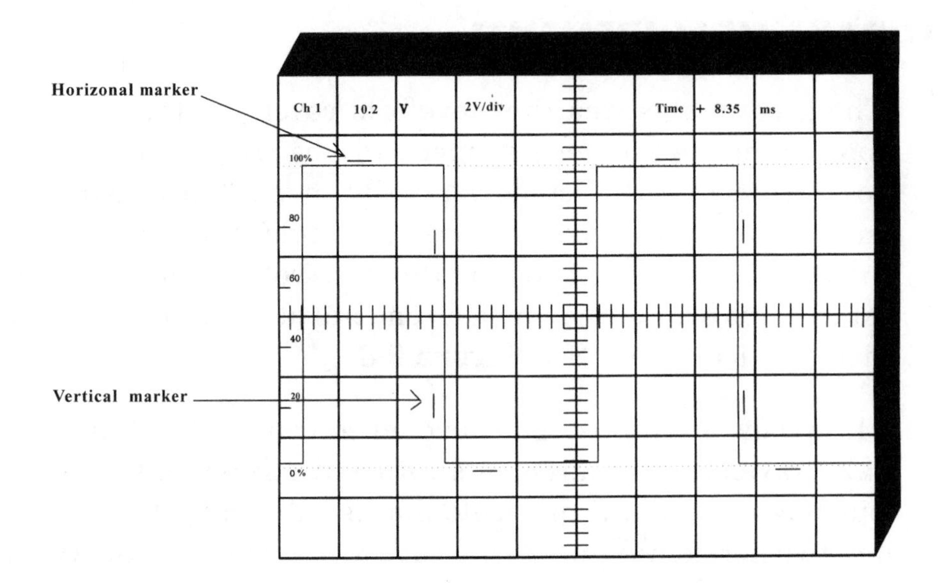

Figure 8-4. Diagram of horizontal and vertical cursor marks.

TIME AND FREQUENCY MEASUREMENTS

Taking time measurements can be tricky if you don't adjust the signal portion of the trace to cover a large area of the screen. When you take time measurements you use the horizontal scale of the oscilloscope. Time measurements include measuring the period, pulse width and pulse timing. In Chapter two we said that the frequency is equal to the reciprocal of the period, so once you know the period of a signal, the frequency is equal to one divided by the period. Like voltage measurements, time measurements are more accurate when you adjust the portion of the signal to be measured to cover a large area of the screen.

Taking time measurements along the center horizontal graticule line, having smaller divisions, makes for more accurate time or frequency measurements. See **Figure 8-5**.

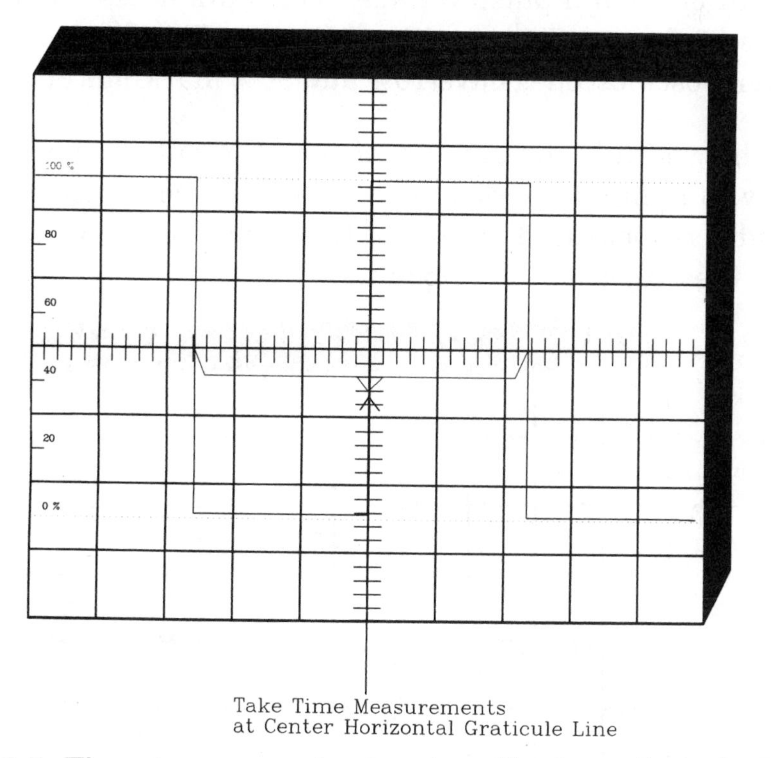

Figure 8-5. Time measurements at center of horizontal graticule line.

RISE TIME & PULSE MEASUREMENTS

Digital circuits are more prone to malfunction. Due to their pulses becoming distorted, digital electronics are more likely to produce errors or become inoperative than most other types of electronics. Because of the fine detail of their pulse shapes, the shape of pulses used in digital electronics is very important. When pulses become distorted, the timing in the pulse train is often changed significantly enough to cause timing errors in electronic equipment.

Standard pulse measurements are *pulse width* and *pulse time*. Rise time is the amount of time a pulse takes to go from the low to high voltage. Rise time is measured from the 10% to 90% point of the full pulse voltage. This eliminates any irregularities at the pulse's transition corners. This is also the reason many oscilloscopes have 10% and 90% markings on the screen.

Pulse width is the amount of time the pulse takes to go from low to high and back to low again. By convention, the pulse width is measured at 50% of the full voltage. See **Figure 8-6** for these measurement points.

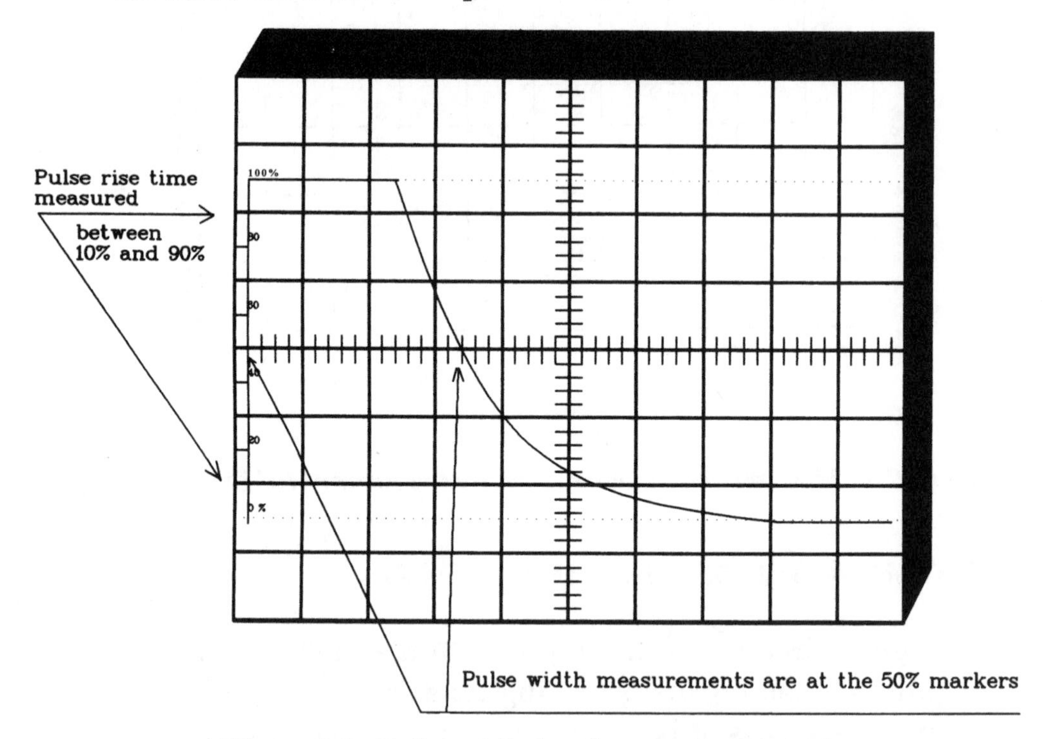

Figure 8-6. Pulse and rise time measurements.

Pulse measurements often require fine tuning of the triggering. Those oscilloscopes with horizontal magnification greatly helps in measuring fine details of fast pulses. To become an expert at capturing pulses you must learn by doing.

PHASE MEASUREMENTS

Its is said that a picture is worth a thousand words. On most oscilloscopes the horizontal control sections have an XY mode or provide two input channels labeled Y-input and the other X-input. The XY mode lets you display an input signal rather than the time base function on the horizontal axis. This mode of operation opens up a whole new area of phase measurement techniques.

As you know from reading Chapter 2, the phase of a wave is the amount of time that passes from the start of one cycle to the beginning of the next cycle, measured in degrees. Phase shift describes the difference in timing between to otherwise identical periodic signals.

A French physicist named Jules Antoine Lissajous (pronounced LEE-saz hoo) named waveforms resulting from XY arrangements as *Lissajous patterns*. The Lissajous pattern is created when one sine wave signal is fed into the input of vertical channel Y and another sine wave signal is fed into the other input vertical channel X. **Figure 8-7** shows how to set up an oscilloscope in the XY mode.

When using the XY mode both signals must be a sine wave. This measuring technique is called an XY measurement because both the X and Y axis are tracing voltages. From the shape of the Lissajous pattern, you can tell the phase difference between the two signals. You can also tell their frequency ratio. **Figure 8-8** shows Lissajous patterns for various frequency ratios and phase shifts.

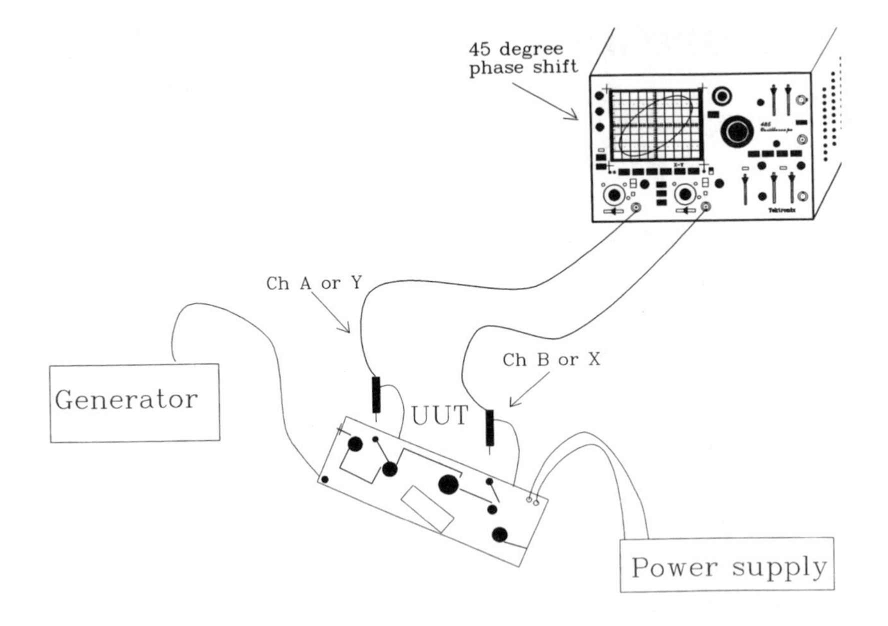

Figure 8-7. Diagram of XY mode setup.

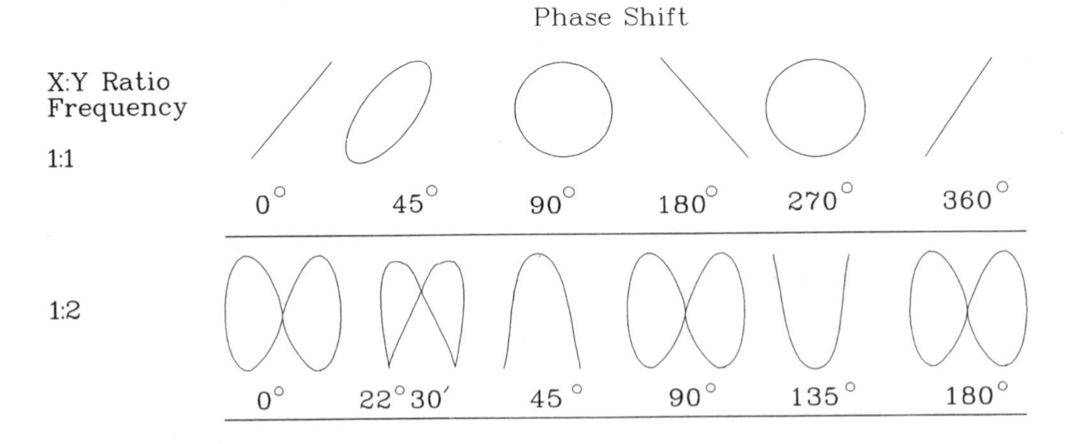

Figure 8-8. Lissajous patterns of two sine waves.

This chapter has covered basic measuring techniques that will help you when using your oscilloscope. The measuring technique you use will depend on your application. In the next two chapters we will explore more powerful measuring techniques.

SUMMARY

The measuring technique you use will depend on your specific application.

Voltage and time measurements are more accurate when measured in the center of the screen and cover a large area of the screen.

Time measurements are more accurate when measured in the center of the screen and the trace covers a large area of screen space.

The small square divisions on the screen graticule are called *major divisions.*

Tick marks on the horizontal and vertical graticule are called *minor divisions.*

The percent (%) markers are used for rise time measurements.

Ohm's law states that voltage is equal to current times resistance.

AC power is more complicated to calculate than DC power.

Power is equal to voltage times current.

Digital circuits malfunction more because of distortion in pulse shapes and pulse timing.

phase shift and frequency ratio of two sine waves can be determined from the Lissajous patterns.

Phase shift describes the difference in timing between two identical periodic signals.

Pulse measurements often require fine turning of the oscilloscope's triggering.

CHAPTER 8 QUIZ

1. The percentage markers are used to measure what type waveform measurements?

   A. Pulse
   B. Analog
   C. Digital
   D. None of the above

2. The square divisions on the graticule face are?

   A. Minor divisions
   B. Level changes
   C. Major divisions
   D. Time lines

3. Ohm's law states that the current is equal to_____?

   A. E÷R
   B. IxR
   C. E÷I
   D. None of the above

4. Voltage ÷ resistance is equal to_____?

   A. Current B. Resistance
   C. Voltage D. Impedance

5. Electrical power is equal to voltage x resistance?

   A. True B. False

6. For the most accurate voltage measurements use a small screen area.

   A. True B. False

7. In AC power measurements, RMS values are used in the power formula.

   A. True B. False

8. Frequency is the reciprocal of time.

   A. True B. False

9. The period of a waveform should be measured at the center of the X and Y axis on the CRT.

   A. True B. False

10. What type of electronic device will malfunction if the pulse shape is distorted?

    A. Analog device
    B. RF device
    C. Digital device
    D. None of the above

11. The amount of time a pulse takes to go from low to high is the ________?

    A. Frequency
    B. Rise time
    C. Mode
    D. None of the above

12. Pulse measurements require fine-tuning of what control on the oscilloscope?

    A. Vertical control
    B. Horizontal control
    C. Trigger control
    D. Time/Div control

13. When both of the X and Y input channels are used in the XY mode, the waveform that results is ________?

    A. Current waveform
    B. HV waveform
    C. Lissajous waveform
    D. Negative waveform

14. Jules A. Lissajous was a German physicist.

    A. True
    B. False

15. XY mode can also be used for measuring eye patterns.

    A. True
    B. False

NOTES

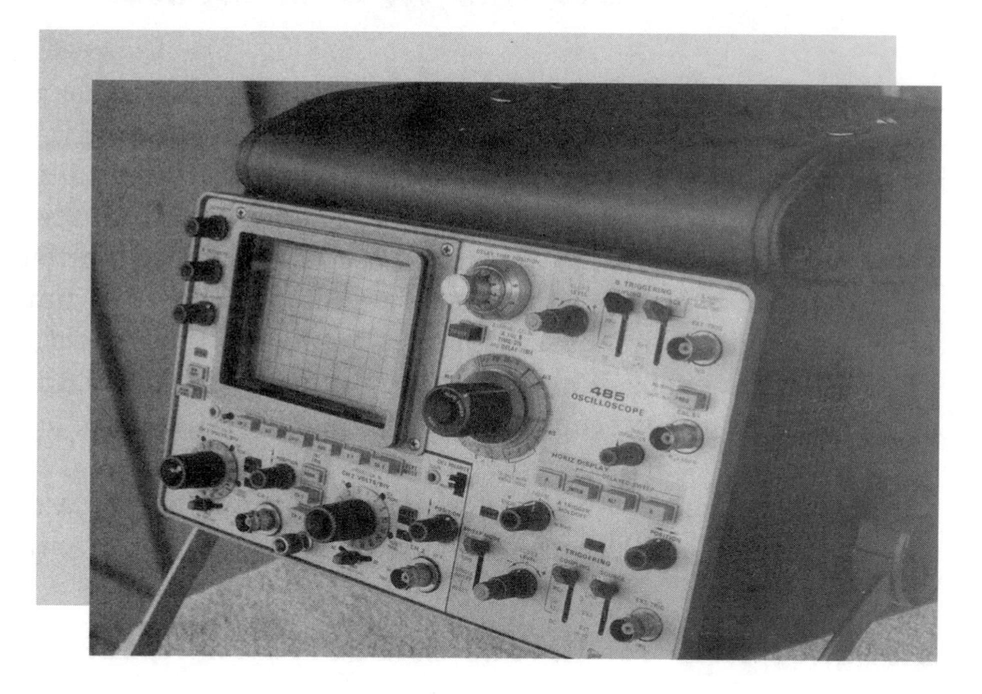

CHAPTER 9
DIFFERENTIAL MEASUREMENT

CONTENTS:

The oscilloscope depends on a signal voltage and/or current signal from the test circuit in order to create a waveform of meaningful data for the user. In contrast to the digital multimeter (DMM), todays oscilloscopes are unsurpassed in their measuring capabilities. When an oscilloscope uses a differential probe it transforms the oscilloscope into a super instrument for displaying the more subtle measurements that would just fly over the knobs of most digital multimeters. This chapter will describe many techniques used for making differential measurements. It will also explore the different types of measurements you can make with a differential probe, and how to use a conventional oscilloscope when measuring differential voltages.

DIFFERENTIAL VS. SINGLE-ENDED TECHNIQUES

When an oscilloscope is equipped with a differential amplifier, or uses a differential probe for making measurements, the oscilloscope becomes a test tool for troubleshooting problems that are impossible to detect with conventional single-ended measurements.

Single-ended measurements often hide high power distribution problems because the oscilloscope's ground lead provides an alternate ground path from which the signal is measured. In addition to the measurements changing, single-ended measurements often alter the circuits operation.

When using a differential probe it is possible to track down problems not normally seen via conventional measurements. For example, placing the inputs of a differential probe right on an IC's power supply lead gives a true picture of the device's power condition. Even if the power supply looks clean, both the ground and power pins may be moving with respect to each other's grounds in the system. By moving the differential probe, it is possible to view ground gradients between an individual device ground and other grounds in the system. Digital bounce is a good example of a ground gradient. Placing a differential probe on the inputs and ground pins of a device gives a true picture of what signals that device actually *sees* at its inputs and ground.

A differential amplifier amplifies the difference of two signals when applied to the amplifier inputs, and rejects any voltage which is common to both inputs. See **Figure 9-1**. The transfer equation is:

$$Vo = Ap\,(V_{+in} - V_{-in})$$

Where:

Vo is referenced to earth ground.

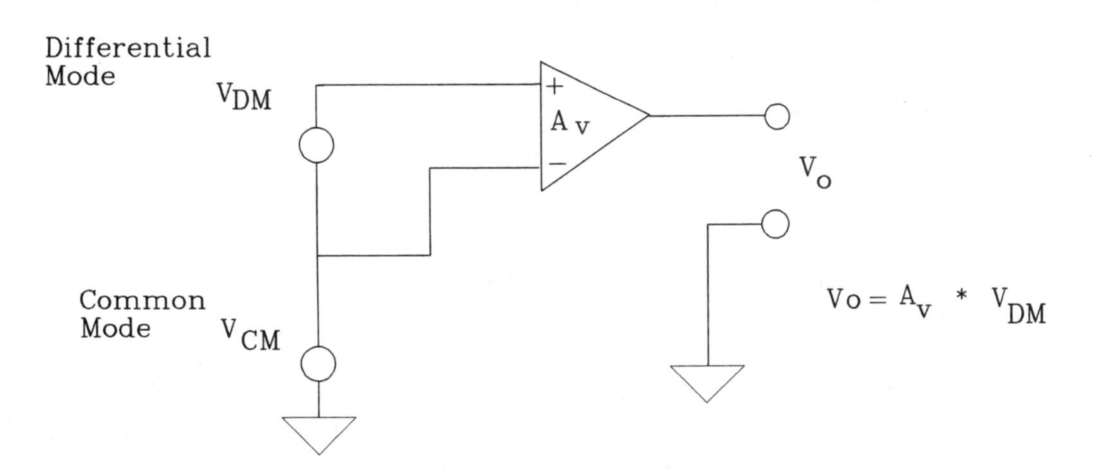

Figure 9-1. Diagram of differential amplifier with input signals.

The voltage of interest, or difference signal, is referred to as the differential or differential mode signal and is expressed as V_{DM} (V_{DM} is the V_{+in} - V_{-in} term in the transfer equation above). **Figure 9-2** is a diagram of a differential amplifier making a floating measurement.

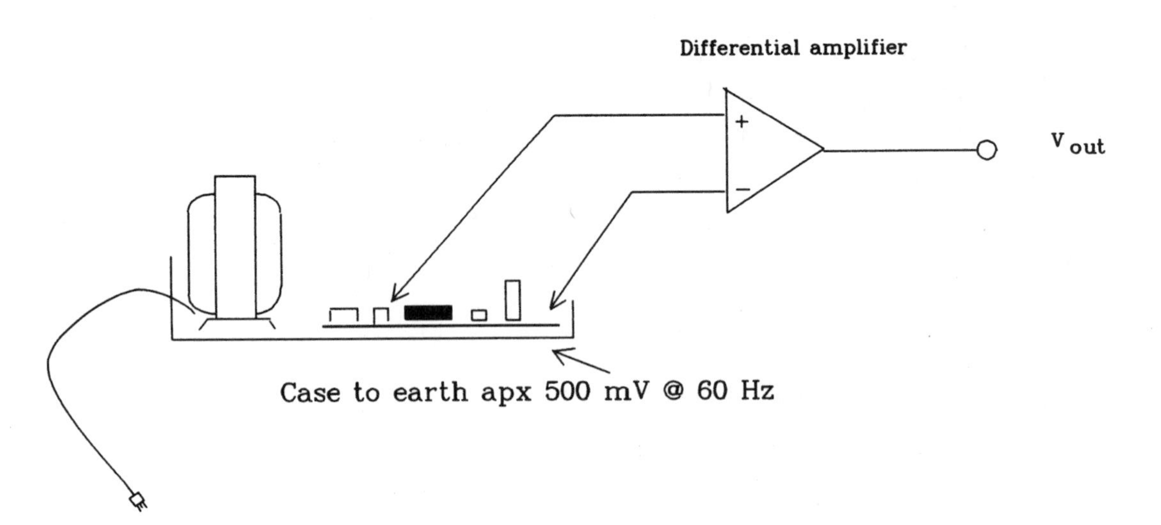

Figure 9-2. Differential amplifier measuring on a floating ground.

The voltage which is common to both inputs is referred to as the *Common-Mode Voltage* expressed as V_{CM}. The characteristic of a differential amplifier to ignore the V_{CM} is referred to as *Common-Mode Rejection* or CMR. The ideal differential amplifier rejects all of the common-mode component, regardless of its amplitude and frequency.

In **Figure 9-3**, a differential amplifier is used to measure the collector/emitter voltage at TP1 and TP2 in the step-down switching power supply. As the control element switches from ON to OFF the emitter voltage swings positive, then negative. The collector swings to zero volts and then to a positive voltage which is output to the catch diode. The differential amplifier allows the oscilloscope to measure the true V_{CE} signal at a sufficient resolution such as 2V/division while rejecting most of the unregulated output voltage source and ground.

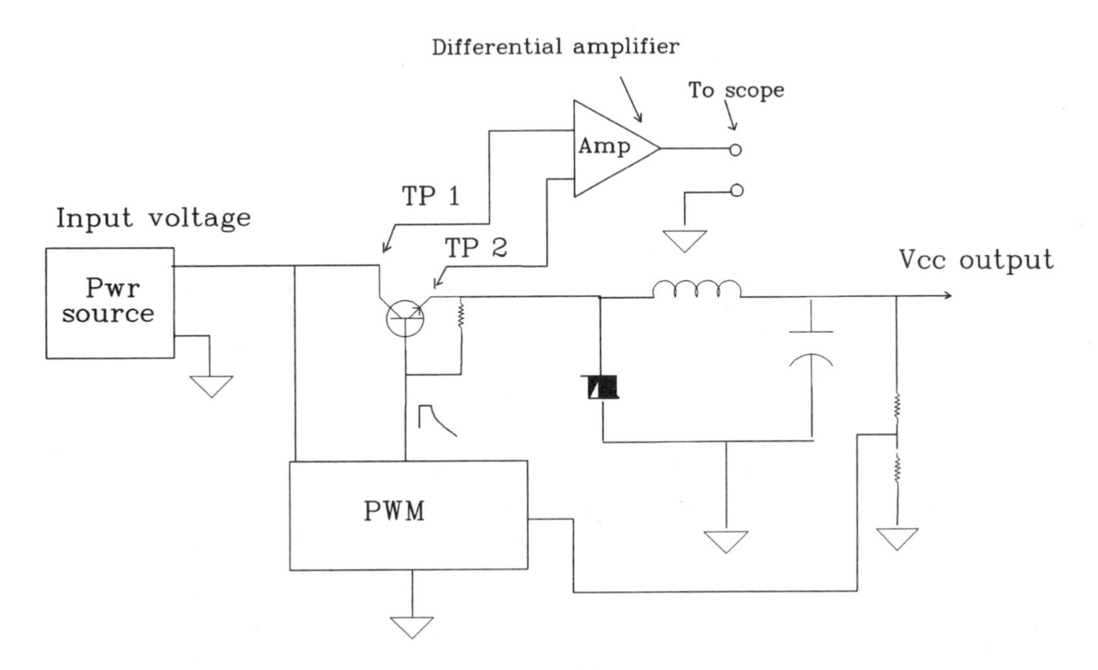

Figure 9-3. Switching supply connected to a differential amplifier.

DIFFERENTIAL AMPLIFIERS

Common-Mode Rejection Ratio (CMRR) is the ability of differential amplifiers to eliminate the undesirable common-mode signal. All differential amplifiers have a small amount of common-mode voltage which appears as an error signal in the output, making it indistinguishable from the desired differential signal. There are two mathematical formulas used to describe CMRR. The first is "square root of the signal to noise ratio of the output divided by the signal to noise ratio of the input." The second, and most common, is defined as "differential-mode gain divided by common-mode gain referred to the input."

$$CMRR = \sqrt{\frac{SNR_{output}}{SNR_{input}}}$$

$$CMRR = \frac{A_{DM}}{A_{CM}}$$

For evaluation purposes, we can assess CMRR performance with no input signal. The CMRR then becomes the apparent V_{DM} seen at the output resulting from common-mode input. It is expressed either as a voltage ratio or in dB:

$$Db = 20\log\frac{A_{DM}}{A_{CM}}$$

To get a better idea of the differential amplifier's CMRR let's examine the following example. Suppose we need to measure the voltage in the output damping resistor of an audio power amplifier, as shown in **Figure 9-4**. In this example a CMRR of 100,000:1 would be equivalent to 100 dB. At full load, the voltage across the damper (V_{DM}) should reach 35 mV, with an output swing (V_{CM}) of 80 Vp-p.

The differential amplifier we use has a CMRR specification of 100,000 at 1kHz (with the amplifier driven to full power with a 1 kHz sine wave, one hundred thousandth or 100 dBs at 1 kHz). With the amplifier driven to full power with a 1 kHz sine wave, one hundred thousandth of the common-mode signal will erroneously appear as V_{DM} at the output of the differential amplifier, which would be 80V/100,000 or 800 μV. The 800 μV represents up to a 2.3% error in the true 35 mV signal.

The measured output voltage could be as high as 35.6 mV with a 2.3% error. The CMRR specification is an absolute value, and does not specify polarity (or degrees of phase shift) of the error. Therefore, the user cannot simply subtract the error from the displayed waveform. CMRR is generally the highest (best) at DC and degrades with increasing frequency of the V_{CM}. Some differential amplifiers plot the CMRR specification as a function of frequency.

If we were to try to measure the CMRR error in **Figure 9-4**, we would quickly run into problems. The common-mode switching signal is a square wave and the CMRR specification assumes a sinusoidal (sine wave) component. Because the square wave contains energy at frequencies considerably higher than 30 kHz, the CMRR will probably be worse than specified at the 30 kHz point.

TEST AND MEASUREMENT

Because the common-mode component expects a sine wave, a simple test exists that can quickly determine the extent of the CMRR error. See **Figure 9-5**.

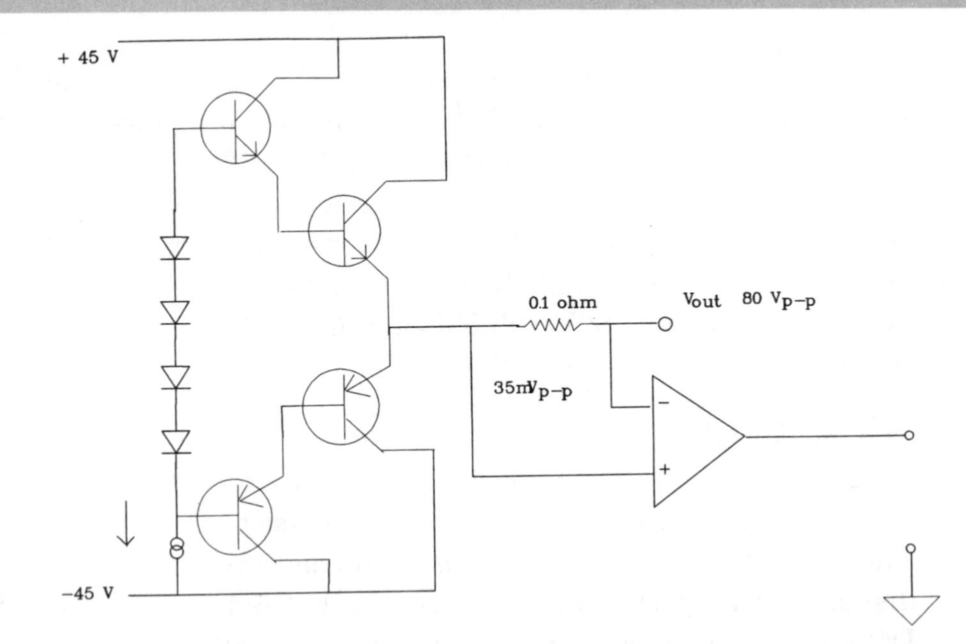

Figure 9-4. Common-mode error from a differential amp with 100,000 CMRR.

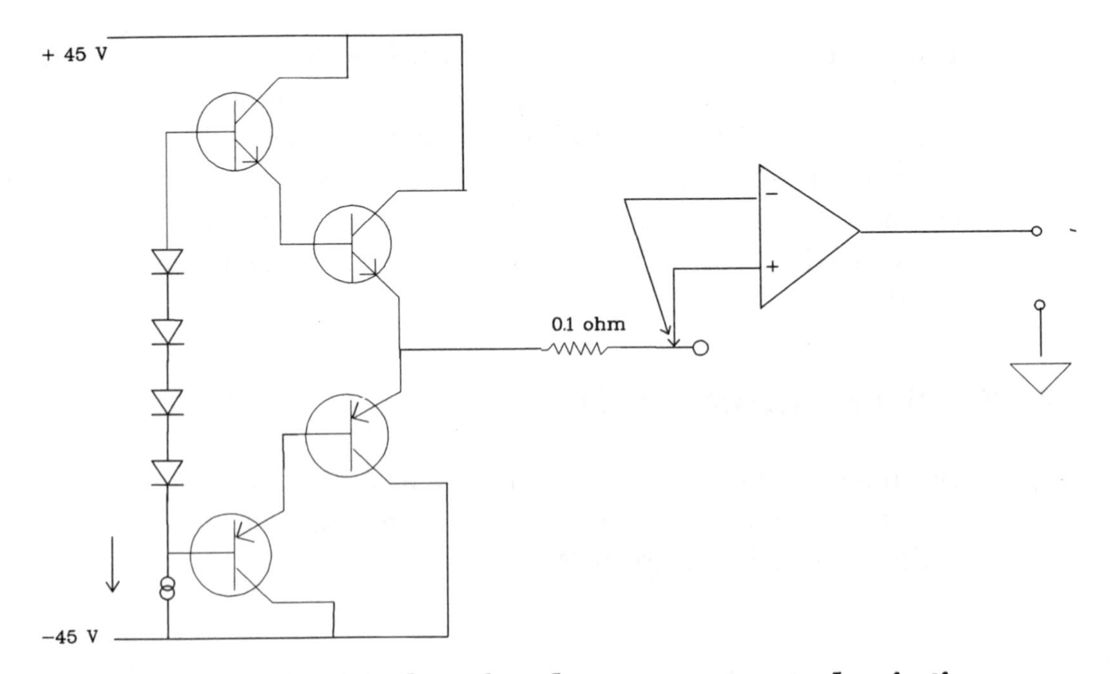

Figure 9-5. Empirical test for adequate common-mode rejection.

Temporarily connect both input leads to the signal source (in our example, the emitter/resister **Figure 9-5**). The oscilloscope is displaying only the common-mode error. You can now determine if the magnitude of the error signal is significant.

Remember, the phase difference V_{CM} and V_{DM} is not specified. Therefore, subtracting the displayed common-mode error from the differential measurement will not accurately cancel the error.

The test is outlined for determining the extent of the common-mode rejection error in the actual measurement environment. However, there is one effect this test will not catch. With both inputs connected to the same point, there is no difference in driving impedance as seen by the amplifier.

This situation produces the best CMRR performance. When the two inputs of a differential amplifier are driven from significantly different source impedance, the CMRR will be degraded.

DIFFERENTIAL AND ISOLATOR PARAMETERS

In order to take full advantage of the many uses and applications of differential amplifiers and probes, the user must become acquainted with some of the more general specifications and parameters of theses devices. An understanding of differential amplifiers and differential probes will help you choose the best measuring technique for your application. Of these specifications the following two outlined here are of special importance.

THE DIFFERENTIAL MODE RANGE

The differential mode range is equivalent to the input range specification of an amplifier or single-ended oscilloscope input. Input voltage which exceeds this range will overdrive the amplifier, resulting in output clipping or non-linearly.

Common-mode range is the voltage window over which the amplifier can reject the common-mode signal. The common-mode range is usually larger than or equal to the differential range. Depending on the amplifier topology, the common-mode range may not change with different amplifier gain settings. Exceeding an amplifier's common-mode range may have various results in the output. In some situations, the output will not clip and may produce a close approximation of the true input, with some additional offset. In this situation, the display may be close enough to what is expected that it is not questioned by the user. It is always a good practice to verify that the common-mode signal is within the acceptable common-mode range before making any differential measurements.

MAXIMUM COMMON-MODE SLEW RATE

Maximum common-mode slew rate is specified for some differential amplifiers and most isolators. This specification is often confusing but very important. Part of the confusion results from a lack of standard definitions between instrument manufacturers. Also, differential amplifiers and isolators behave differently when their maximum common-mode slew rate is exceeded. Essentially, maximum CM slew rate is a supplemental specification to CMRR. The specification is usually given in units of kV/ms.

Some types of differential amplifiers, like other amplifiers, reach a large signal slew limitation before the small-signal bandwidth specification is exceeded. When one or both sides of a differential amplifier are driven to slew-rate limiting, the common-mode rejection is degraded very rapidly. Unlike CMRR, maximum slew rate does not imply an increasing amount of common-mode feed-through in the output. Once the maximum common-mode slew rate is exceeded, all bets are off—the output is likely to clamp at one of the power supply rails.

ISOLATOR PARAMETERS AND LIMITATIONS

In isolators, however, the effect is more gradual - like CMR in a differential amplifier. As the common-mode slew rate increases (as opposed to frequency), more of the common-mode component "feeds through" to the output. Intuitively, the specification would imply a maximum slew rate at which a known amount of feed-through appears in the output. It is important to note that with some isolators, the CV slew rate specification is actually a maximum nondestructive limit. The ability to make meaningful measurements is lost at slew rates much lower than the maximum specification. When using an isolator it is best to test the common-mode feed-through before making critical measurements. This is easily done by driving both the probe tips and the reference lead with the same common-mode signal and observing the output. **Figure 9-6** shows a diagram of an isolation probe.

QUASI-DIFFERENTIAL TECHNIQUES

Conventional single-ended oscilloscopes cannot make differential measurements without using special probes or isolators to protect the circuit and instrument from damage. Attempting to

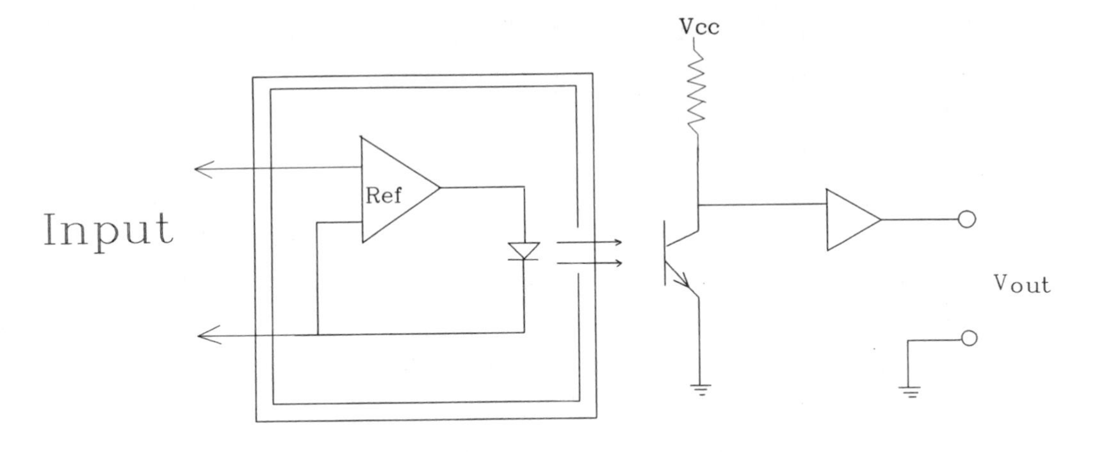

Figure 9-6. Diagram of isolation probe.

do so without any of these devices will only cause a small fireworks display. However, most single-ended oscilloscopes do have the ability to make simple differential measurements. This mode of oscilloscope use is called *quasi-differential.*

While limited in performance, this technique may be adequate for some measurements. To make a differential measurement, two vertical channels are used, one for the positive input and the other channel for the negative input. See **Figure 9-7**. The channel used for the negative must be set to invert mode and the display mode set to "ADD channel A + channel B".

For proper operation, both inputs must be set to the same scale factor, and both input probes must be identical models. The display now shows the difference in voltage between the two inputs.

To maximize the CMRR, the gain in both channels should be matched. This can easily be done by connecting both probes to a square wave source with an amplitude within the dynamic range of the Volts/Div setting (about +6 divisions). Set one of

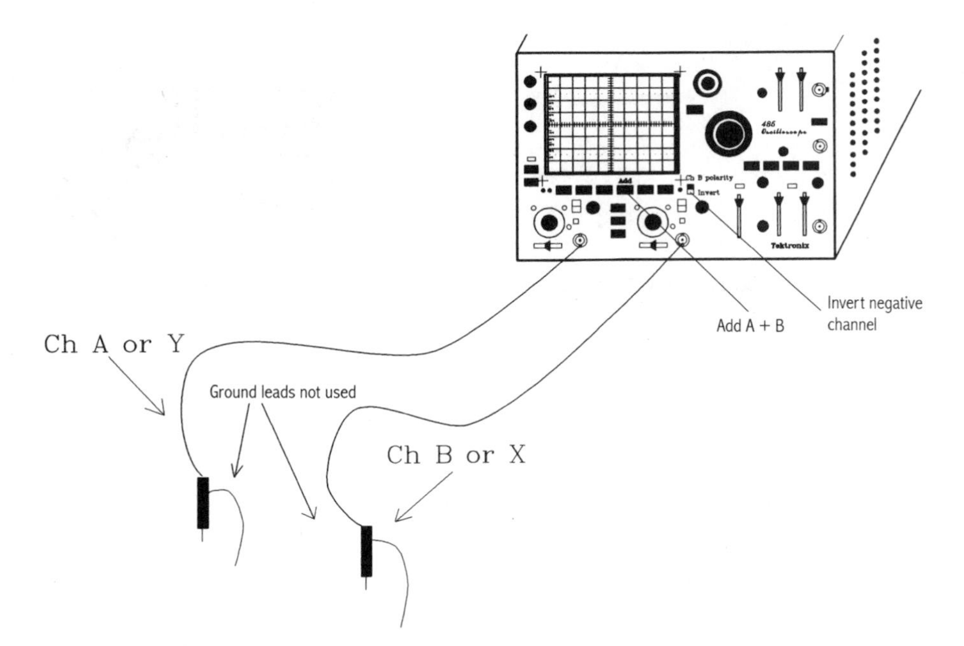

Figure 9-7. Quasi-differential mode and setup.

the channels to "uncalibrated-variable" gain and adjust the variable gain control until the displayed waveform becomes a flat trace. See **Figure 9-8**.

The primary limitation of using this technique is a rather small common-mode range, which results from the oscilloscope's vertical channel dynamic range. Generally, this is less than ten times the Volts/Div setting from ground. Whenever $V_{CM} > V_{DM}$, this mode of obtaining a differential result can be thought of as extracting the small difference from the two voltages.

Most digital storage oscilloscopes perform waveform math in the digital domain, after the analog signal has been digitized. The limited resolution of the analog-to-digital converter is often not adequate to view the resulting differential signal after the common-mode signal is subtracted out. But the AC gain in the two channels is not precisely matched, as CMRR at higher frequencies is rather poor.

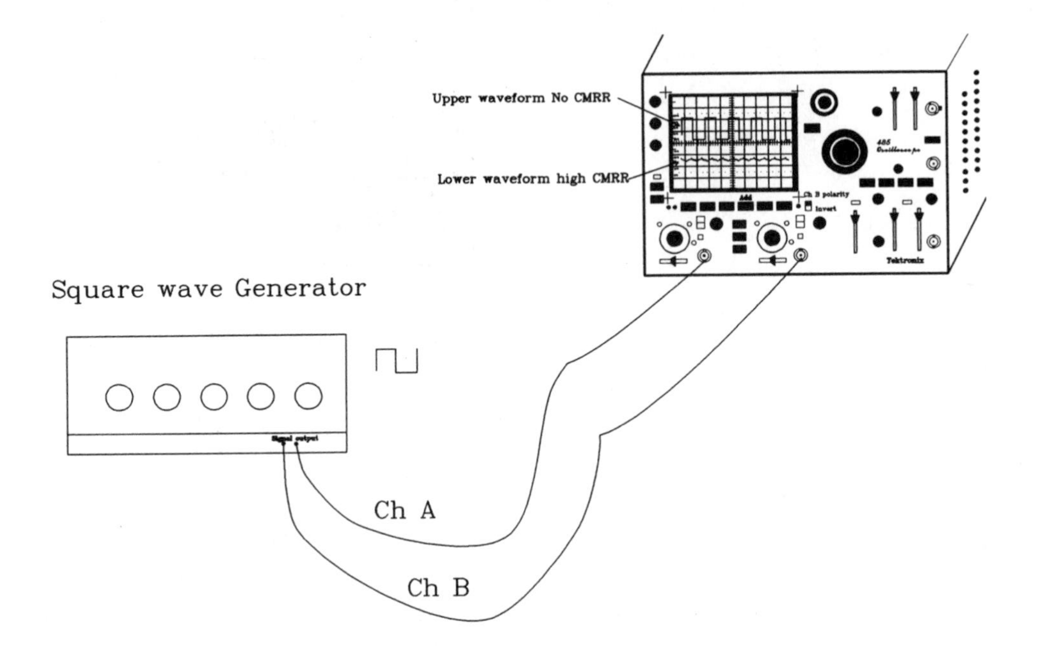

Figure 9-8. Oscilloscope inputs to square wave generator maximizes CMRR in both input channels for quasi-differential mode.

This technique is suitable for applications where the common-mode signal is the same or lower amplitude than the differential signal, and the common-mode component is DC or low frequency, such as 50 or 60 Hz power line. It effectively eliminates ground loops when measuring signal of moderate amplitude.

HIGH-VOLTAGE DIFFERENTIAL PROBES

A new probe has appeared on the market. A high-voltage active differential probe. Its new topology, which boasts of new fixed attenuation with switchable differential gain, allows these probes to keep their full common-mode range in gain settings. The single attenuator greatly reduces complexity resulting in lower cost to the user.

These types of probes provide an affordable, safe method of measuring line-connected circuits commonly found in switching power supplies, power inverters, motor drives, electronic lamp ballasts, etc. With common-mode ranges up to 1,000V, these probes eliminate the need for the extremely dangerous practice of floating the oscilloscope. Recently, workplace hazard monitoring organizations such as OSHA (Occupational Safety and Health Administration) have intensified their verification of equipment grounding, issuing costly fines to violators.

In addition to safety benefits, the use of these probes can improve measurement quality. An obvious benefit is the full use of the oscilloscope's multiple channels with the simultaneous viewing of multiple signals referenced to different voltages.

Because the probes are true differential, both of the inputs are high-impedance, high-resistance, and low capacitance. Floating oscilloscopes and isolators do not have balanced inputs. The reference side (the ground clip on the probe) has a significant capacitance to ground. Any source impedance the reference is connected to will be loaded during fast common-mode transitions, attenuating the signal.

Worse yet, the high capacitance can damage some circuits. See **Figure 9-9**. Connecting the scope common to the upper gate in an inverter may slow the gate-drive signal, preventing the device from turning off and destroying the input bridge. This failure is usually accompanied with a bright spark and some smoke at your test bench, something many technicians can attest to.

With the balanced low input capacitance of high-voltage differential probes, any point in the circuit can be safely probed with either lead.

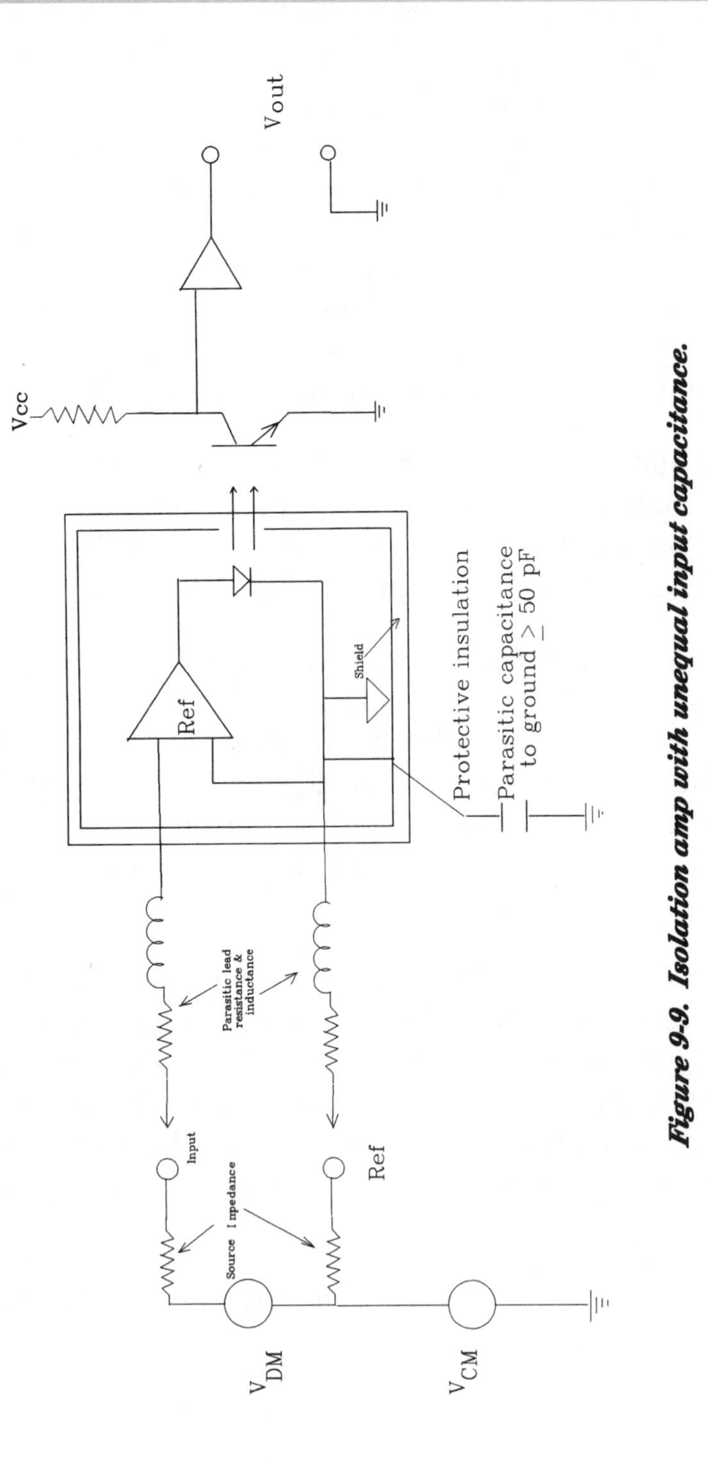

Figure 9-9. Isolation amp with unequal input capacitance.

SUMMARY

Differential measurement enables the user to make safe and accurate floating measurements.

Conventional oscilloscopes have the capabilities to make differential measurements.

CMRR is a measure of a differential amplifier's ability to eliminate undesirable signals.

CMRR can be expressed as either a voltage ratio or in dBs.

Differential-mode range is equivalent to the input voltage range of an amplifier or oscilloscope input.

Isolators are also used to make differential measurements.

Quasi-differential measurements can be made with a conventional dual-channel oscilloscope.

An ideal differential amplifier amplifies the difference of the two signals and totally rejects any voltage which is common to both inputs.

CHAPTER 9 QUIZ

1. An ideal differential rejects any voltage on the inputs?.

   A. True B. False

2. CMRR measures _______ on a differential amplifier.

   A. Common no mode
   B. Common-mode rejection
   C. Noise rejection
   D. None of the above

3. A voltage gain of 10,000 is what in dBs _______?

   A. 60 dB
   B. 80 dB
   C. 50 dB
   D. 75 dB

4. A differential probe is never used for measuring floating voltages.

   A. True
   B. False

5. Common-mode range refers to ______?

   A. Range of voltage
   B. Rejection of the CMS
   C. Voltage window
   D. None of the above

6. Quasi-differential measurements refer to ______?

   A. Using scope inputs
   B. Half differential
   C. Non-isolated
   D. None of the above

7. What do the letters OSHA stand for?

   A. Occupational Safety and Health Administration
   B. Oscilloscope Safety Hazard Activists
   C. Occupational Safety Hazard Act
   D. None of the above

8. High-voltage differential probes are what type of probe?

   A. Passive
   B. Logic
   C. Active
   D. Temperature

9. All oscilloscopes have built-in differential amplifiers.

   A. True
   B. False

10. Differential and isolator are the same type of probe.

    A. True
    B. False

11. Newer types of digital oscilloscopes have a math mode function built in.

    A. True
    B. False

12. Digital oscilloscopes have limited resolution in viewing a differential waveform.

    A. True B. False

13. CMRR is better at higher than lower frequencies.

    A. True B. False

14. A flat trace is the result of adjusting ______ on a conventional oscilloscope when maximizing the CMR.

    A. Gain controls
    B. Brightness controls
    C. Height controls
    D. Volt/Div controls

15. To maximize CMRR the gain in both channel must be ______.

    A. Uneven
    B. Greater than 1
    C. Matched
    D. Less than 1

NOTES

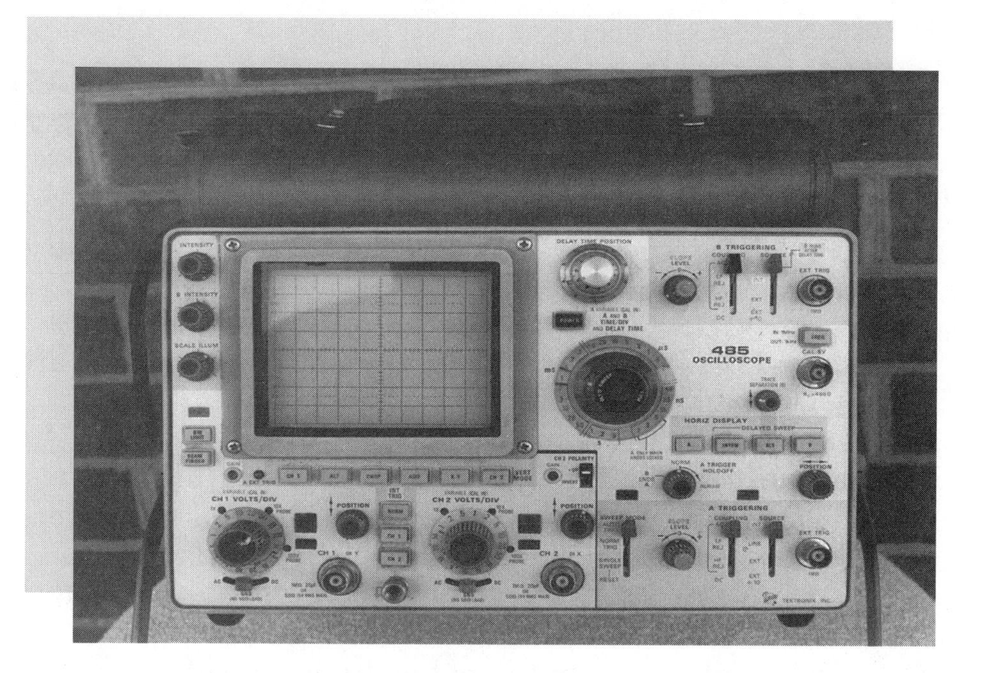

APPENDIX A

ARTICLES:

Power-Electronics Measurements

Measuring Instantaneous Power

Measuring True/Apparent Line Power

Measuring Instantaneous AC Resistance

Measuring Line-voltage Distortion

Low-level Current Measurements

Monitoring Core Saturation

POWER-ELECTRONICS MEASUREMENTS MADE EASY WITH TDS OSCILLOSCOPES: MEASURING BATTERY POWER CONSUMPTION

Problem: There are no convenient methods to measure power consumption in battery-powered devices.

Key TDS Features: Waveform multiplication, and area measurements.

Benefits: The TDS waveform processing functions also apply to slow DC events often associated with chart recorders.

A critical specification of all portable electronic gadgets is battery life. It's easy to measure the steady-state DC power consumption with a traditional multimeter - just measure the DC current drain and battery voltage and multiply the two results. But most electronic devices switch between a multitude of operating modes. Hence, you need to measure the dynamic power characteristics of the various modes over time to truly understand system performance.

Figure 10 shows various power measurements for the "rewind" mode of a portable tape player. The top two traces are the battery voltage and current. The TDS calculates the power waveform by multiplying the voltage and the current. The waveform signatures are easily understood. The battery voltage dips as the motor starts drawing current. The current surges to start the motor. The power remains relatively constant until the rewinding finishes; then the motor stalls and the current surges again. The motor then shuts off.

The TDS calculates the area under the current waveform as 20.7 A-sec. This is converted to 5.8 mA-h by dividing by 3600 s/h You can then compare this result to the mA-h capacity of your selected battery type to estimate operating times.

On the other hand, if you choose to calculate true energy consumption, you need to multiply the battery voltage and current, then measure the area under the power curve. The TDS calculates the instantaneous power waveform (400 mW/div) and measures the area under the curve. In this case the result is 58 W-s or 58 joules. While this seems like a large number, it's equivalent to running a 60 Watt light bulb for about one second! Finally, the TDS calculates the width of the power curve which measures the total event time of just over two minutes.

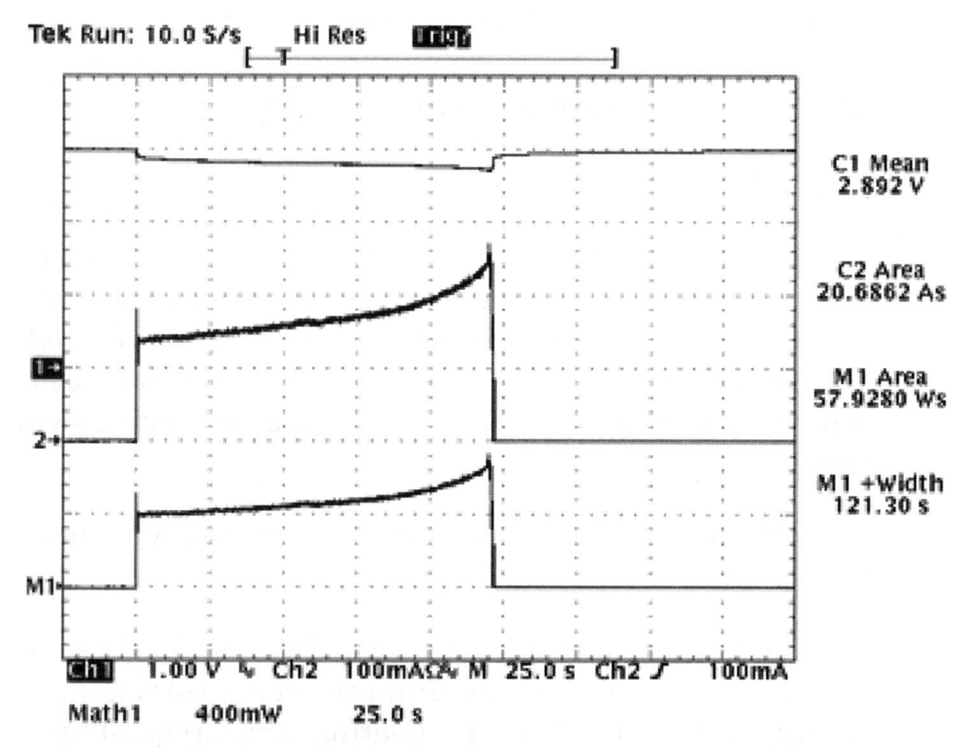

Figure 10. Upper waveform is 3 V battery voltage. Center waveform is the battery current (100 mA/div). Lower waveform is the calculated instantaneous power waveform. The width of the power waveform measures the total rewind time of 121 seconds.

MEASURING INSTANTANEOUS POWER IN SWITCHING TRANSISTORS

Problem: Displaying only the voltage and current waveforms in a switch-mode transistor doesn't characterize the instantaneous power dissipation in the device.

Key TDS Features: Waveform multiplication and automatic measurements.

Benefits: The TDS waveform processing and display capabilities readily translate the raw voltage and current waveforms into more meaningful power switching parameters.

Proper selection of switch-mode transistors in power converters requires a detailed understanding of both the steady-state and transient modes of operation. In particular, the start-up mode often places the greatest stress on devices. Dynamic measurements include verification of Vgs or Vbe drive levels and corresponding Vds or Vce transitions. But the most useful dynamic measurement is the power dissipation during switching transitions.

In particular, this is the relationship between Vds and I d or between Vce and I c.

Figure 18 illustrates the start-up cycle of a 40 kHz converter. The upper waveform is the drain-to-source voltage measured through the P5205. This is a floating measurement since neither terminal is at ground. The TCP202 senses the drain current at 500 mA/div. The TDS waveform multiplication function was used to generate the lower waveform representing instantaneous power. The displayed vertical scale factor for the power waveform is 25 W/div. The TDS calculates a peak power of 30 W during this initial turn-on transition. This was by far the worst case dissipation for the device. For example, the initial turn-off transition power near the end of the waveform was negligible.

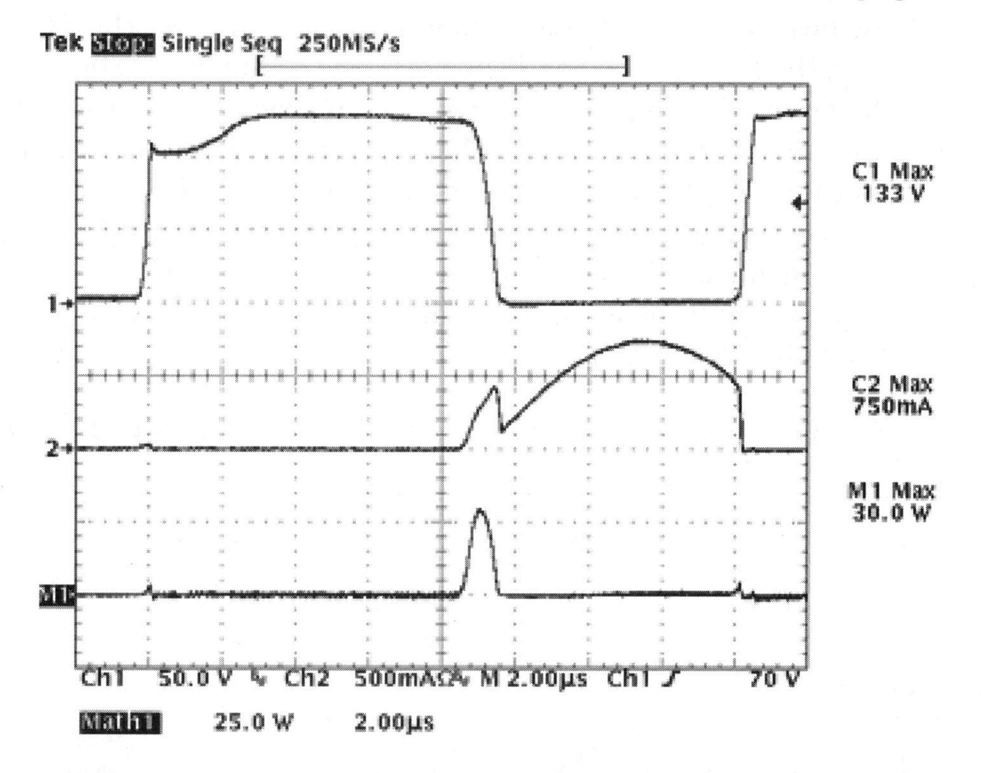

Figure 18. The upper waveform is the floating drain-to-source voltage measured with the P5205. The center waveform is the drain current measured with the TCP202 (500 mA/div). The TDS multiplication function generates the lower waveform showing the instantaneous power (25 W/div).

Calculating the instantaneous power waveform in switch-mode converters is a classic example of the big number/small number relationship. The blocking voltage is large during cut-off when no current flows. Conversely, there's a small saturation voltage when the device conducts the maximum current.

This means that any small DC offsets in the voltage or current measurements can lead to gross numerical errors. Typical symptoms of offset problems include excessive power during cut-off or negative power measurements. When you use the TCP202 and the P5205 with the TDS, perform the following simple step before any instantaneous power measurement. Set the TDS to measure the mean of both the voltage and current waveforms. Set the TCP202 and P5205 to the sensitivity settings that will be used for the measurement. Then, with the device-under-test turned off, use the DC offset adjustment on the TCP202 and the P5205 to null the mean level for each waveform. Always be sure that all three instruments have passed their warm-up interval.

MEASURING TRUE AND APPARENT LINE POWER

Problem: Erroneous results in AC power measurements due to inadequate sample rate and record length.

Key TDS Features: Decimal record length, waveform multiplication, and automatic measurements.

Benefits: The TDS decimal record length simplifies the measurement of line-power parameters.

Quantifying AC line-power components is a basic requirement in every off-line power converter application. The digital scope is an ideal tool for this task, but some subtle traps can turn this straightforward measurement into a tedious exercise. The first step is to acquire the voltage and current waveforms (Figure 1). A standard 10X passive probe can safely sense the line voltage.1 The TCP202 can readily sense the line current. Figure 2 shows the results for a typical switch-mode converter.

The upper waveforms show the distorted current and sinusoidal voltage waveforms. The TDS calculates the individual RMS values of 121 Volts and 1.11 Amps. The apparent power is 134 VA or the product of the individual RMS values. True power is the mean value of the instantaneous product of voltage and current. The TDS multiplies the voltage and current waveforms to create the instantaneous power waveform. The TDS then calculates the mean value of the entire power waveform which yields a true power of 88 W.

The ratio between the true and apparent power values yields the power factor of 0.66.

Apparent Power (PA) =
120.8 V * 1.108 A =
133.8 W

True Power = (Pt) = 88.0 W

Power Factor (pf) =
88.0 W = 0.66
133.8 W

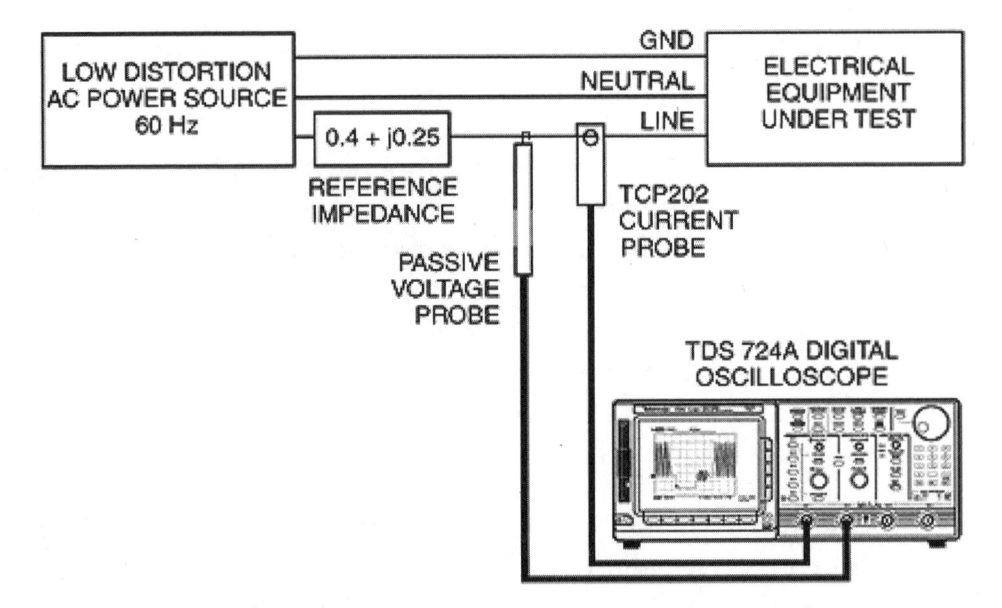

Figure 1. A standard 10X passive probe senses the line voltage while the TCP202 current probe clips on to the line lead. The scope implicitly provides the ground return for the voltage probe. NOTE: Do not hook the probe's ground clip to the neutral wire.

Several system features simplify this technique as well as guard against measurement errors. For current sensing, the TCP202 is ideal because it provides a flat frequency response across the relevant measurement range.

Contrast this with AC current transformer (CT) clamps. Many CT devices are not suitable for sensing the complex waveshapes at switching converter inputs. While rated for 50 or 60 Hz operation, their low-frequency roll-off can induce phase shift with respect to the voltage measurement. This phase shift will cause errors in the true power measurement since the two waveforms are multiplied in time. It's important to note that CT devices will very accurately sense the current's fundamental frequency component, but they're not necessarily designed to sense the higher frequency harmonics (up to several kHz) found in contemporary power waveforms.

Several TDS features also simplify this technique. As with any measurement of a periodic signal, a rule is to record and measure over complete events.

In the case of AC waveforms, this means selecting a measurement interval that includes an integral number of line cycles. With some planning, you can set your TDS to capture integral numbers of 60 Hz (or 50 Hz) cycles. In Figure 2, the waveforms are exactly 50 milliseconds long or three complete line cycles. This results from a sampling rate of 50 kS/s (i.e., 20 μs between samples) and a record length of 2500 samples. Unlike the TDS, many digital scopes only offer record lengths that are powers of 2 such as 1024 samples. When combined with a limited selection of sample rates (e.g., 10 kS/s, 20 kS/s), this yields recording times that don't capture an integral number of line cycles. For example, a sample rate of 20 kS/s and a record length of 1024 samples results in 3.07 line cycles (instead of 3.00 cycles), which translates to a calculation error of greater than 2%.

In some cases, you can obtain acceptable results by using a digital scope's cycle-based measurement functions. The scope scans the selected waveform for a complete cycle of data and performs the measurement only on a complete cycle of the waveform. This technique works well with simple sinusoidal signals such as measuring the RMS value of the line voltage, but it can lead to erratic results with complex current and true power waveforms. In addition, a single cycle measurement of the 120 Hz instantaneous power waveform, shown in Figure 2, only represents half of a 60 Hz line cycle.

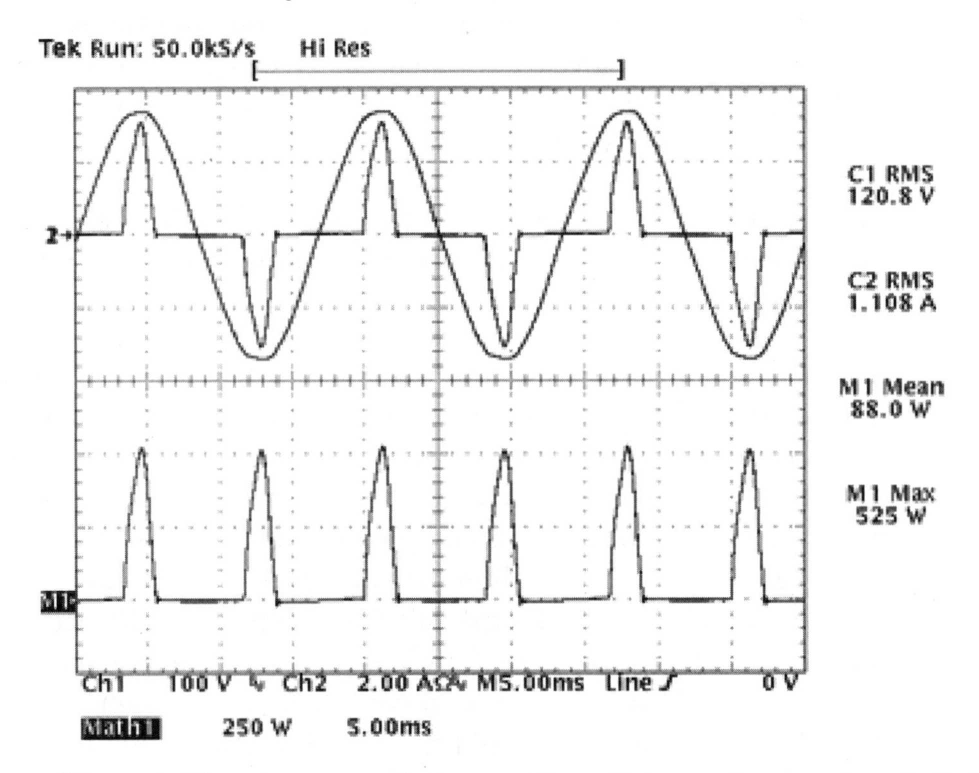

Figure 2. The upper waveforms are the voltage and current scaled at 100 V per division and 2 A per division. The TDS multiplies the voltage and current to create the instantaneous power waveform (lower) scaled at 250 W per division. The TDS then calculates the power parameters. The true power is 88 W and the peak power is 525 W.

Of course there are applications where you want to directly control the measurement interval. You may want to measure the difference between the power delivered on the two halves of the line cycle or test your converter's power consumption at 47 Hz using a programmable AC source. In these cases, you would use the TDS gated measurement capability to directly set the time interval for a calculation. This capability is illustrated separately in this note.

[1] This assumes your AC neutral and ground are at the same potential. If your building's ground wiring has a "power-quality" problem, or if you need to actually measure the line-to-neutral voltage, use the P5205 differential probe. Do not connect the ground clip of your standard 10X passive probe to the AC neutral. Under no circumstances should you "float" your scope by cutting or defeating the ground connection. For further information, please read the publication 51W-10640-0, "Floating Oscilloscope Measurements... And Operator Protection."

MEASURING INSTANTANEOUS AC RESISTANCE

Problem: In AC signals, a relevant parameter is often measurable only at aspecific portion of the line cycle. It's difficult to interpret acycle-location dependent parameter over many cycles of a continuous AC waveform.

Key TDS Features: Waveform division, FastFrame acquisition, and gated measurements.

Benefits: The FastFrame acquisition mode simplifies analysis of complex AC events by capturing only selected segments of successive cycles.

Several types of power electronics components exhibit time-varying resistance. For example, current limiting devices exhibit large resistance changes by design as a function of temperature variations due to resistive heating. The time constant associated with the variations may be a few seconds or even less than a second. Multimeters don't work because you can't use the meter's ohms mode with a live circuit. And while you could independently measure and divide the AC voltage and AC current levels through the component, you can't obtain dynamic information.

At first glance, the conventional digital scope might be the solution. By capturing both the AC voltage and AC current waveforms, you can divide the two waveforms and obtain the instantaneous resistance. This is shown in Figure 11 for the best known time-varying AC resistance - the incandescent filament. The two sinusoidal waveforms are the AC voltage (larger) and AC current; the TCP202 was set to 1A/div. The third waveform is the derived instantaneous resistance calculated using the TDS waveform division function. The problem is obvious. At the zero crossings of the AC waveform, the current goes to zero. This leads to a divide-by-zero condition twice per cycle. While this by itself is not a problem, it makes interpretation much more difficult. In this case, we're only looking at about one line cycle. But if we had captured ten cycles, the display would be cluttered with these excursions to ±*.

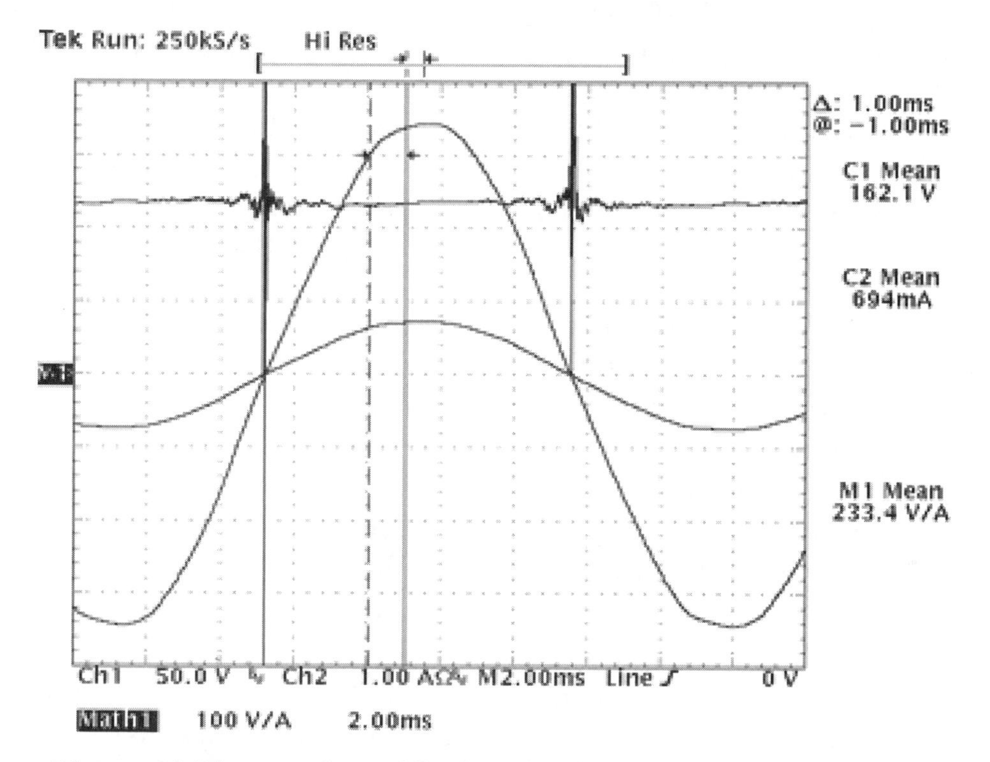

Figure 11. The two sinusoids show the line voltage and current to an incandescent filament. The TCP202 was set to 1 A/div. The third waveform is the instantaneous resistance calculated by dividing the voltage waveform by the current waveform (100 V/A or 100 Ohm per div). The large excursions near the zero crossings result from the divide-by-zero condition. The cursors define a gated measurement interval of 1.0 milliseconds.Figure 12. FastFrame was used to capture 50 point segments at the positive line peaks. Each segment was defined to start when the line voltage reached 150 V. Since the sampling rate was 1 MS/s, each segment represents a 50 microsecond sliver of a total line cycle. The twenty segments shown, requiring 1,000 total points,

represent the information portion of a real-time interval of over 330 milliseconds (20 cycles / 60 Hz). The same information captured without FastFrame would have required a record length of 330,000 points.

The TDS gated measurement function provides one approach to extracting the relevant information. We can choose to measure the voltage and current waveforms within a user-settable zone defined by the vertical cursors. In this case, we're near the positive peak of the cycle and the mean voltage (Ch 1) in the zone is 162 V and the mean current (Ch 2) is 694 mA. The TDS calculates the mean value of the divided waveform in the same zone as 233 Ohm. Of course, you could obtain the same results by manually dividing the mean voltage and current values.

But this somewhat tedious process is greatly simplified by the TDS 500's FastFrame capability. Recall that the objective is to capture information only when it's valid. In this case, we'd ideally capture snapshots of the voltage, current, and resistance waveforms at the line peaks. Figure 12 shows the result for the lamp turn-on interval. The TDS was set to capture 50 μs snapshots when the line voltage reached 150 V. The sampling is then "disabled" until exactly one cycle or 16.7 ms later. The voltage waveform now appears as a straight line with a level of ~150 V. The current snapshots change rapidly during the first few cycles and settle out to 642 mA. Since the voltage is constant in this example, the resistance waveform is proportional to the reciprocal of the current waveform and settles to a steady-state value of 237 Ohm in less than 10 line cycles. You can still use the TDS gated measurement capability to scroll through the start-up waveform and measure values at different points in time. Of course, you need to keep in mind that each 50 μs line segment or frame is separated by one line cycle period.

One way to think of FastFrame is as a record-length multiplier. Power electronics is filled with measurement applications where the relevant information occurs at well defined instants of consecutive-line or switch-mode cycles. Rather than filling digital scope memory with extraneous information, FastFrame lets you apply your waveform memory only when it's needed.

MEASURING LINE-VOLTAGE DISTORTION

Problem: Visual inspection of simple parametric relationships such as crest factor are inadequate for line-voltage distortion.

Key TDS Features: FFT, automatic measurements, and cursor readout.

Benefits: The FFT can quantify distortion levels that are masked by traditional measurement techniques.

The companion measurement to line-current distortion is line-voltage distortion. This is often overlooked because the voltage waveform generally appears to be a textbook sinusoid. But manufacturers of AC inverters or uninterruptible power sources (UPS) need to characterize the output characteristics of their equipment. Once again, a task that used to require specialized distortion-analysis equipment has become more accessible through the digital processing capabilities of the TDS.

The lower waveform in Figure 4 shows the line voltage. The TDS calculates the 180 V peak and 128 V RMS values of the waveform. The crest-factor ratio is 1.42, which matches that of an ideal sinusoid. But the FFT results on the upper waveform uncover much more information. Unlike the current-harmonic measurement, the harmonic content of the voltage waveform is much smaller. In this case, it's more appropriate to set the TDS to display the results in dBV or on a log scale where 1 V (RMS) equals 0 dBV.

You can convert between a value V(x) in Volts and its equivalent dBV(x) in dBV using:

$$V(x) = 10(dBV(x) / 20)$$

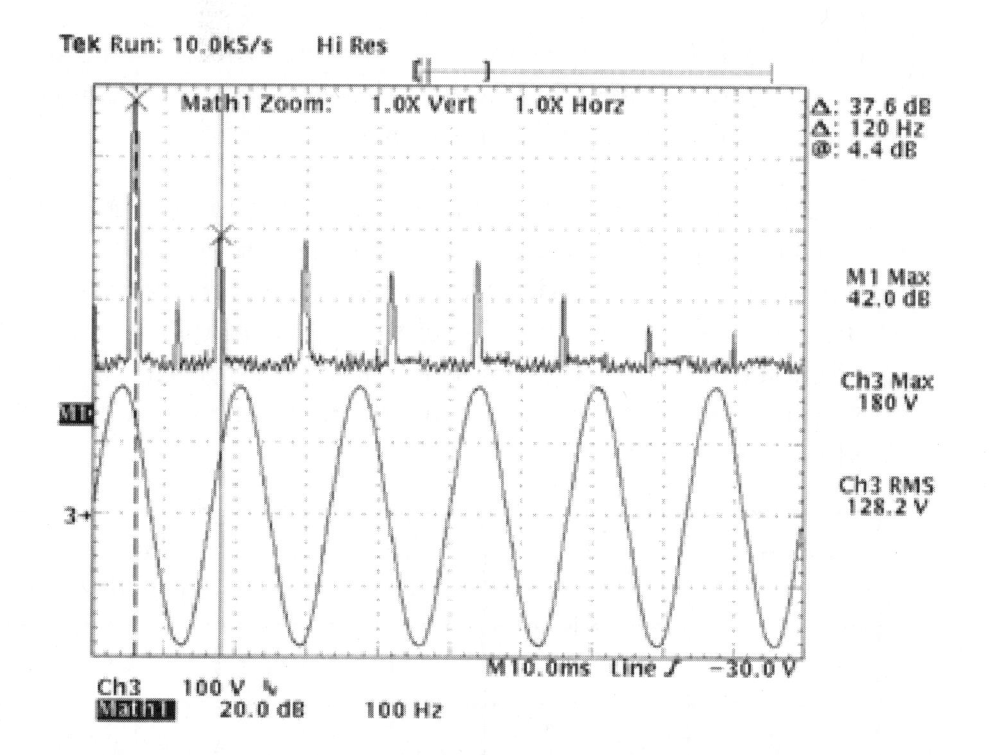

Figure 4. The lower waveform is the line voltage. The maximum and RMS voltages are auto- matically calculated at the right. The upper wave- dBV(x) = 20 log V(x) form is the FFT of the voltage waveform. The vertical scale is 20 dBV/div and the horizontal scale is 100 Hz/ div. The principal component is the 60 Hz fundamental at 42.0 dBV (126 V).

The peak harmonic component is the 60 Hz fundamental at 42.0 dBV or 126 V (i.e., 102.1). The TDS cursor function lets you quickly scroll through the harmonics to measure relative levels. For example, the 3rd harmonic, which is 120 Hz from the fundamental, is 4.4 dBV (1.7 V) and is 37.6 dB below the fundamental. You can insert -37.6 dB into the formula above (yielding 0.013) to conclude that this component is 1.3% of the fundamental. Table 1 summarizes the results. You can calculate the total harmonic distortion (THD) by taking the square root of the sum of the squares of each value in the last column. In this case the THD is 1.9%. For line voltages, the total distortion is typically dominated by the first few odd harmonics so you only need to tabulate three or four values.

The TDS cursor function can directly display the level of each harmonic relative to the fundamental. This simplifies THD calculations since distortion components are normalized to the fundamental amplitude. THD is obtained by taking the square root of the sum of the squares of the last column. In this case, the total is 1.9%, but the 3rd and 5th harmonic alone would have yielded a THD of 1.8%.

Harmonic	Frequency	Relative to Fundamental	Percent of Fundamental
2nd	120 Hz	-56.4 dB	0.15%
3rd	180 Hz	-37.6 dB	1.32%
5th	300 Hz	-38.8 dB	1.15%
7th	420 Hz	-47.2 dB	0.44%
9th	540 Hz	-45.2 dB	0.55%
11th	660 Hz	-54.8 dB	0.18%

Table 1

LOW-LEVEL CURRENT MEASUREMENTS

Problem: The usefulness of digital multimeters is limited as they provide only static measurement results.

Key TDS Features: Hi-Res acquisition and zoomed display.

Benefits: The stability and signal processing capability in the TCP202/TDS combination provide a new way to characterize low-level current dynamics.

The goal of increasing battery life in handheld electronic gadgets requires thorough knowledge of how current is drawn by the load components.

Microampere current measurements are readily made using digital multimeters, but they don't provide more than a steady-state or DC measure of current flow. How do you measure micro-power switching dynamics as loads switch from active to shutdown states?

The TCP202 provides a simple technique for multiplying the sensitivity of current measurements in this application. By looping 'n' turns of the sensed-current conductor through the current probe (Figure 8), you can multiply the effective sensitivity by n. The waveform in Figure 9 (lower screen) is the battery-current waveform of a clock sensed at the maximum TCP202 sensitivity of 10 mA/div. However, the sensed current conductor was looped through the current probe ten times, so the sensitivity is actually ten times larger (1 mA/div). The TDS calculates the mean multiplied current of 620 μA which is equivalent to 62 μA. Note that the current pulses occur once per second as the clock advances the second hand. Many multimeters cannot provide a meaningful current measurement in this application. In addition, the TDS zoom function can expand the pulse (upper screen) to reveal the current dynamics. Note that the TDS displays a box around the area that's magnified into the upper screen.

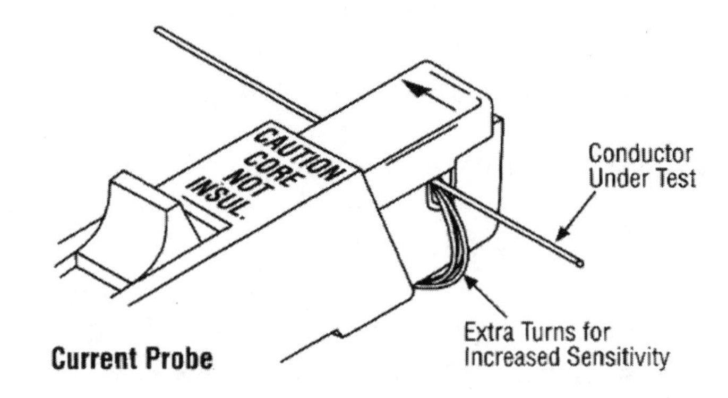

Figure 8. Looping 'n' turns of the sensed-current conductor through the current probe multiplies the effective sensitivity by n.

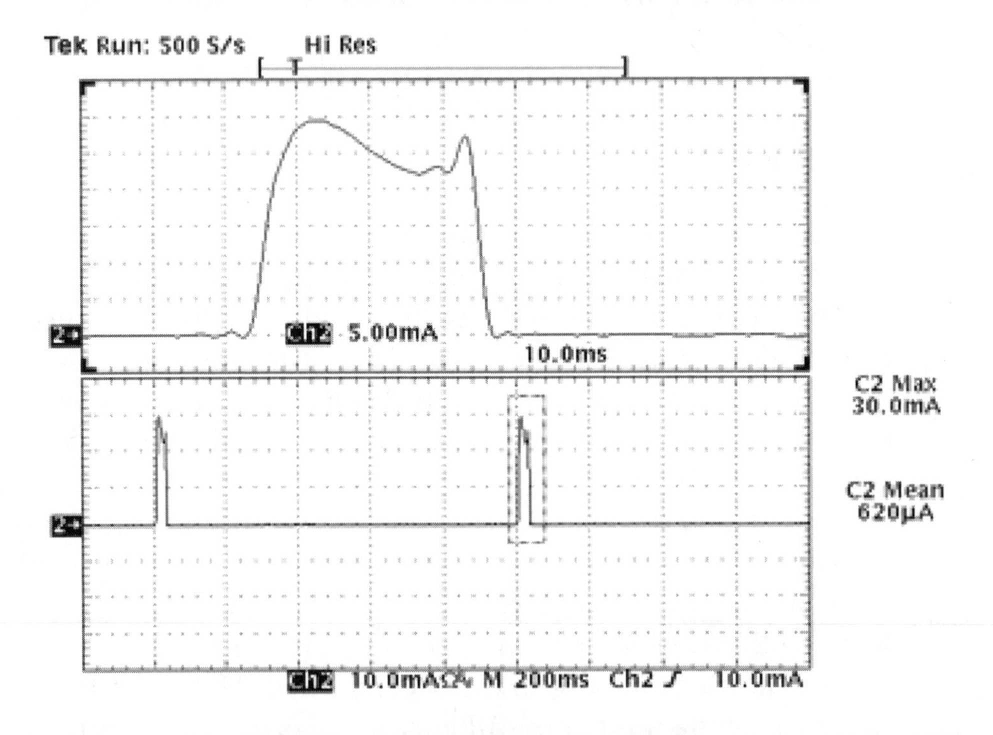

Figure 9. The TCP202 was set to its maximum sensitivity of 10 mA/div, but the sensed current was looped through the current probe ten times.

Therefore, the TDS calculation of 30 mA (peak) and 620 μA (mean) must be scaled by 1/10. Since the period of the event is one second, the record interval of two seconds captures an integral number of events. Thus, the mean measurement function over the entire record interval can be used to determine the average DC current.

You can readily duplicate these results by following these recommendations.

Since you're taking advantage of the excellent stability of the TDS and TCP202 amplifiers, be sure that both are calibrated and have passed their warm-up periods. Before powering up the circuit under test, use the TDS mean measurement function to calculate the average value of measured current. With the current probe attached to the multi-turn loop, use the TCP202's DC level adjust to null the mean output voltage to 0 mA. Finally, since the multi-turn technique multiplies the noise as well as the signal, you'll need to use the TDS Hi-Res acquisition mode to filter the waveform.

MONITORING CORE SATURATION

Problem: Display of only the voltage and current waveforms in an inductor do not directly reveal core saturation activity.

Key TDS Features: Waveform integration, X-Y display, and zoom.

Benefits: The TDS waveform processing and display functions enable the direct display of an inductor's cycle-by-cycle saturation activity.

Component analyzers readily characterize the out-of-circuit parameters of magnetic elements such as inductors and transformers. Measuring in-circuit dynamics is more challenging but offers valuable insight for start-up and other transient operating modes. One particularly useful measurement is the relationship between a magnetic core's flux density (B) and its field intensity (H). For example, this parametric relationship graphically indicates magnetic saturation which can lead to excessive current flow and thermal runaway in switching transistors. Measuring this relationship in-circuit helps to verify that the core parameters were properly selected for start-up operation where the likelihood of saturation is generally the greatest.

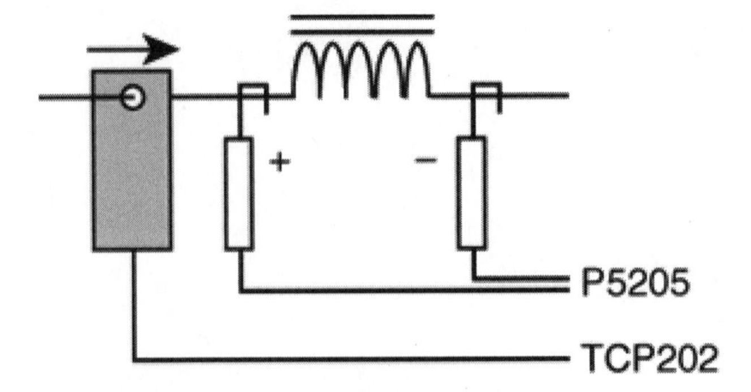

Figure 19. Connecting the P5205 and the TCP202 probes to the inductor under test.

Figure 20 shows the first step of the two-step process for a toroidal inductor during start-up in a 40 kHz converter. The field intensity in the core is directly proportional to the winding current. The TCP202 senses the current (2.84 A pk-pk). The flux density is proportional to the integral of the voltage across the inductor. The P5205 senses the floating voltage since neither side of the voltage is ground. The TDS waveform integration function is used to generate the center waveform which is the time integral of the voltage. Notice from the pk-pk measurement that the units of the integrated waveform are Volt-seconds.

The next step (Figure 21) is to plot the integral of the voltage against the current. The TDS X-Y display mode is used to plot this parametric relationship. The onset of saturation - or the point where flux density no longer increases linearly with field intensity - is now evident. In Figure 20, the '[ ]' brackets above the display show that this X-Y plot is displaying only 20% or 50 μs of the information of the total 250 μs waveform record. In the X-Y mode, you can use the TDS Zoom function to scroll through a reduced time segment of the total waveform. In other words, you can follow the B-H loop as the circuit drives the core.

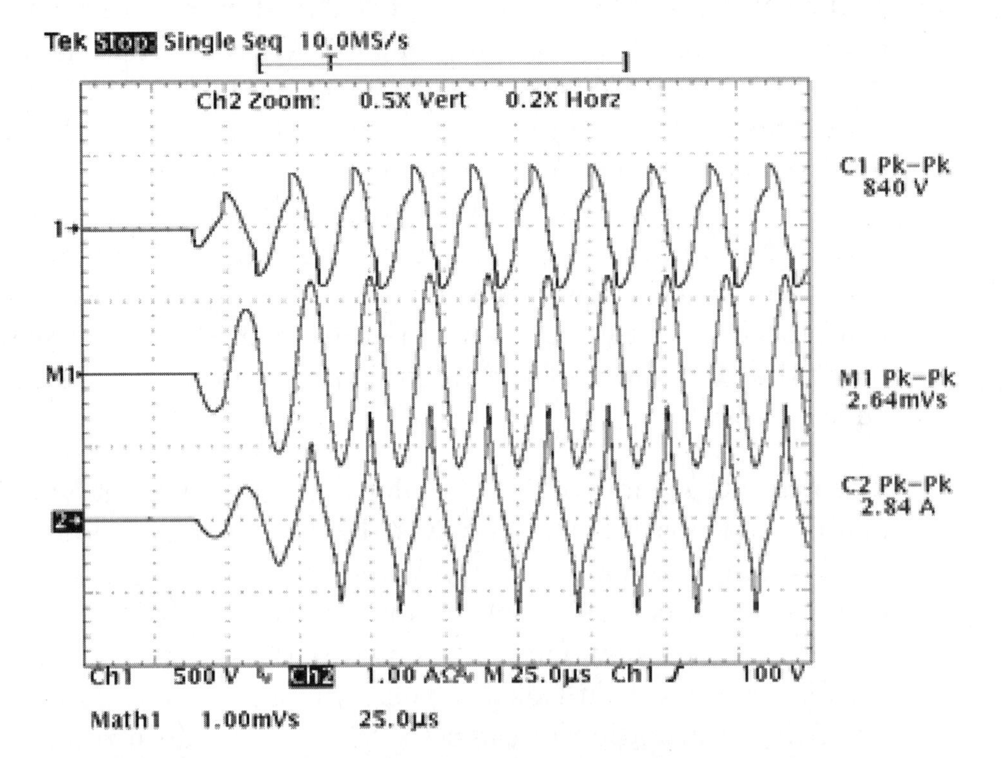

Figure 20. Upper waveform is the floating voltage across the inductor measured with the P5205. The peak-to-peak voltage is 840 V. Center waveform is the integral of the voltage. This is scaled directly with the voltage waveform so the peak-to-peak value is 2.64 mV-sec. Lower waveform is the inductor current. The TCP202 current waveform is displayed at 1.0 A/div.

The peak-to-peak current is 2.84 A. The flux density (integral of the voltage) is plotted against the current. The flux density is 0.5 mV-sec/div, the field intensity is 0.5 A/div. You can use the TDS zoom function to scroll through slices of the start-up waveform to watch how the core enters and exits saturation.

NOTES

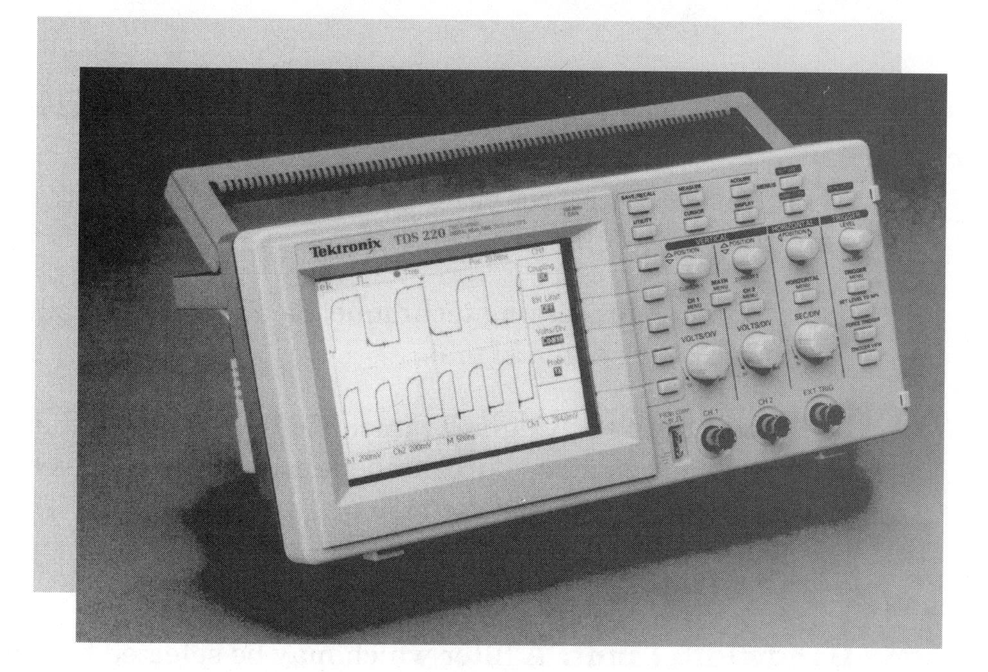

APPENDIX B

GLOSSARY

AC (Alternating Current). A signal in which the and voltage has a repeating pattern over time.

ADC (Analog to Digital Converter). A digital device that converts electrical signals into a discrete binary values.

Alternate Mode. A display mode on an oscilloscope in which traces one channel than switches to the next channel for traces.

Amplitude. The magnitude of a signal. In electronics amplitude usually refers to either voltage or Power.

Attenuation. A decrease in signal level durning its transmission from one point to another.

Averaging. A processing technique used by digital oscilloscope to eliminate noise in a signal.

Balanced Signal. A signal transmitted through a pair of wires, each having the same source impedance. Ground does not service as a return path for the signal.

Bandwidth. A frequency range.

Bandwidth Limit. A filter which may be selected by the user to attenuate noise outside of the bandwidth of interest. Unless otherwise specified, the filer is assumed to be a low-pass topology with a single-pole(-6 dB/octave)roll off.

CRT (Cathode-Ray-Tube). An electron beam tube in which the beam can be focused on a luminescent screen and varied in both position and intensity to produce a visible pattern. A computer terminal display is a CRT.

Chop Mode. A display mode of operation in which small parts of each channel are traced so that more than one waveform can appear on the screen simultaneously.

Circuit Loading. The unintentional interaction of the probe and oscilloscope with the circuit being tested, distorting the signal.

Clamp. A circuit which limits the output swing of an amplifier to keep it within the linear operating range. Usually this is done to reduce the overload recovery time.

Clop. A distorted waveform produced when an amplifier does not have sufficient output voltage range to reproduce the input signal. As the name implies, the output appears as if it was "clipped" off.

Common Mode. The component of the input signal which is common)identical in amplitude and phase) to both inputs of a differential amplifier. An ideal differential amplifier rejects all of the common-mode signal.

Common-Mode Range. The maximum voltage (from ground) of common-mode signal which a differential amplifier can reject. Usually, the common-mode range is greater than the differential-mode range. Depending on amplifier topology, the common-mode range may vary as a function of gain.

Common-Mode Rejection. The elimination of the input common-mode component by a differential amplifier.

Common-Mode Rejection Ratio. The performance measure of a differential amplifier's ability to reject common-mode signals CMRR is expressed as:

CMRR= Differential-Mode Gain/ Common-Mode Gain

Because common-mode rejection generally decreases with increasing frequency, CMRR is usually specified at a particular frequency.

Compensation. A probe adjustment for 10X probes that balances the capacitance of the probe with the capacitance of the oscilloscope.

Coupling. The method of connecting two circuits together. Circuits connection with a wire are directly coupled; circuits connected through a capacitor or a transformer are indirectly (or AC) coupled.

Cursor. An on screem marker that you can align with a waveform to take accurate measurements.

DAC (Digital to Analog Converter). A digital device that converts discrete binary values into electrical signals.

DC (Direct Current). A signal with a constant voltage and current.

Differential Amplifier. A three terminal gain circuit which processes the signal component which is different between two inputs while ignoring the component which is common to the two inputs.

Differential Mode. The signal which is different between the two input inputs of differential amplifier. The differential-mode signal(VDM) can be expressed as:

VDM = (V+input) - (V-input)

Differential-Mode Range. A circuit incorporated in high-gain differential amplifiers to null out DC bias present in the differential input signal. Electrically equivalent to an adjustable battery inserted in series with one of the input leads.

Differential Probe. A probe designed specifically for differential applications. Active differential probes contain a differential amplifier at the probe tip. Passive differential probes are with differential amplifiers and can be calibrated for precisely matching the DC and AC attenuation in both signal paths.

Division. Measurement marking on the CRT graticule of the oscilloscope.

Earth Ground. A conductor that will dissipate large electrical currents into the Earth.

Envelope. The outline of a signal's highest and lowest points acquired over many repetition.

Equivalent Time Sampling. A sampling mode in which the oscilloscope constructs a picture of a repetitive signal by capturing a little bit og information from each repetition.

Floating. A signal which is not referenced to ground. A floating signal cannot be directly measured with a singled ended instrument.

Floating Measurement. A measurement that reads the voltage between two points, neither of which is at ground potential.

Floating the Oscilloscope. The practice of defeating the protective grounding system of an oscilloscope, allowing it to perform floating measurements. Because the entire scope chassis is common to the probe "ground" clip, this dangerous practice my expose the user to electrocution hazards.

Focus. The oscilloscope control that adjusts the CRT electron beams to control the sharpness of the display.

Frequency. The number of times a signal repeats in one second, measured in Hertz(cycles per second). The frequency equals 1/period.

Gigahertz (GHz). 1,000,000,000 Hertz or 10^9 cycles per seconds.

Glitch. An intermittent error in a circuit.

Graticule. The grid lines on a screen for measuring oscilloscope traces.

Ground. A conducting connection by which an electric circuit or equipment is connected to the earth to establish and maintain a reference voltage level. (i.e., the voltage reference point in a circuit.)

Ground Loop. A circuit with multiple low-impedance path connected to the same ground potential. A ground loop acts as a shorted transformer turn which induces circulating ground currents. These currents produce slight changes in the ground potential within the circuit.

Hertz (Hz). One cycle per second; the unit of frequency.

Isolator, Isolated Probe. A device which allows two-point floating voltage measurements to be made with single-ended ground referenced instrumentation. Isolation is accomplished by converting the input signal to optical and/or magnetic (via transformer) form.

Interpolation. A "connect the dots" processing technique to estime what a fast wave looks based on only a few sampled points.

Kilohertz (kHz). 1000 Hertz or 10^3 cycles per second.

Maximum Common-Mode Slew Rate. The upper limit rate of change (dv/dt) of the common-mode component on the input of a differential amplifier or isolator.

Megahertz (MHz). 1,000,000 Hertz or 10^6 cycles per second. A unit of frequency.

Megasamples per second (MS/s). A sample rate unit equal to million or 10^6 samples per second.

Microsecond (μs). A unit of time equal to 0.000001 or 10^{-6} seconds.

Millisecond (ms). A unit of time equal to 0.001 or 10^{-3} seconds.

Nanosecond (ns). A unit of time equal to 0.000000001 or 10^{-9} seconds.

Noise. An unwanted voltage or current in an electrical circuit.

Oscilloscope. An electronic instrument used to display the electrical activity related to the electrical characteristics of a test application.

Peak Detection. An acquisition mode for digital oscilloscopes that lets you see the extremes of a signal.

Peak to Peak (Vpp) Voltage. The voltage measured from the maximum point of a signal to its minimum point, usually equal to 2X the (Vp) level.

Peak Voltage (Vp). The maximim voltage level measured from a zero reference point.

Period. The amount of time it takes a wave to complet one cycle. The period equals 1/frequency.

Phase. The amount of time that passes from the beginning of a cycle to the beginning of the next cycle, measured in degrees.

Pixel. A single sample or picture element in the digital image which is located at specific spatial coordinates.

Probe. An oscilloscope input device, usually having a pointed metal tip for making electrical contact with a circuit element and a flexible cabe for transmitting the signal to the oscilloscope.

Pulse. A common waveform shape that has a fast rising edge, a width, and a fast falling edge.

Quasi Differential. A method of creating a differential amplifier by adding two conventional oscilloscope input channels with one set to invert mode. To produce meaningful results, both channels must be set to the same volts/division position. Compared with true differential amplifiers, quasi-differential mode has very limited common-mode range and lower CMRR., particularly at high frequencies.

RMS. Root mean square. Value of AC voltage.

Real Time Sampling. A sampling mode in which the oscilloscope collects as many samples as it can as the signal occurs.

Record Length. The number of waveform points used to create a record of a signal.

Rise Time. The time taken for the leading edge of a pulse to rise from its minimum to its maximum values (typically measured from 10% to 90% of these values.)

Sample Point. The raw data from an ADC used to calculate waveform points.

Sample Rate. Number of samples per second the ADC can acquire

Screen. The surface of the CRT upon which the visible pattern or waveform is produced.

Signal Generator. A test device for injecting a signal into a circuit input; the circuit's output is then read by an oscilloscope.

Sine Wave. A common curved wave shape that is mathematically defined

Single-Ended. A measurement of a voltage potential referenced to grounded. A conventional oscilloscope input can only make single ended measurements.

Single Shot. A signal measured by an oscilloscope that only occurs once (i.e., transient.)

Single Sweep. A trigger mode for displaying one screenful of a signal and then stopping.

Slide Back (Comparison Voltage). A configuration provided by some differential amplifier which connects a precision calibrated voltage source to one of the amplifier inputs. This provides a single-ended amplifier with an extremely large of calibrated offset. Unlike differential off, Slide back mode can only perform single-ended (round referenced) measurements.

Slope. On a graph or an oscilloscope screen, the ratio of a vertical distance to a horizontal distance. A positive slope increases from left to right, while a negative slope decreases from left to right.

Square Wave. A common wave shape consisting of repeating square pulses.

Sweep. One horizontal pass of an oscilloscope 's electron beam from left to right across the CRT screen.

Sweep Speed. Same as time base. How fast the trace can sweep across the screen

Time Base. Oscilloscope circuitry thats controls the timing of the sweep. The time base is set by the seconds/divsion control.

Trace. The visible shapes drawn on the CRT by the movement of the electron beam.

Transducer. A device that converts a specific physical quantity such as sound, pressure strain, or light intensity into an electrical signal.

Transient. A signal measured by an oscilloscope that only occurs once, aslo called a single-shot event.

Trigger. The circuit that initiates a horizontal sweep on an oscilloscope and determines the beginning point of the waveform.

Trigger Holdoff. A control that inhibits the trigger circuit from looking for trigger level for some specified time after the end of the waveform.

Trigger Level. The voltage level that a trigger source signal must reach before the trigger circuit initiates a sweep.

Volts. The unit of electric potential difference.

Voltage. The difference in electric potential difference.

Waveform. A graphic representation of a voltage varying over time.

Waveform Point. A digital value that represents the voltage of a signal at a specific point in time. Waveform points are calculated from sample points and stored in memory.

X-Axis. The signal in an oscilloscope that controls the beams left to right direction. The X-axis is the axis in which time measurements are made.

Y-Axis. The signal in an oscilloscope that controls the beams up and down direction. The Y-axis is the axis in which amplitude measurements are made.

Z-Axis. The signal in an oscilloscope that controls electron-beam brightness as the trace is formed.

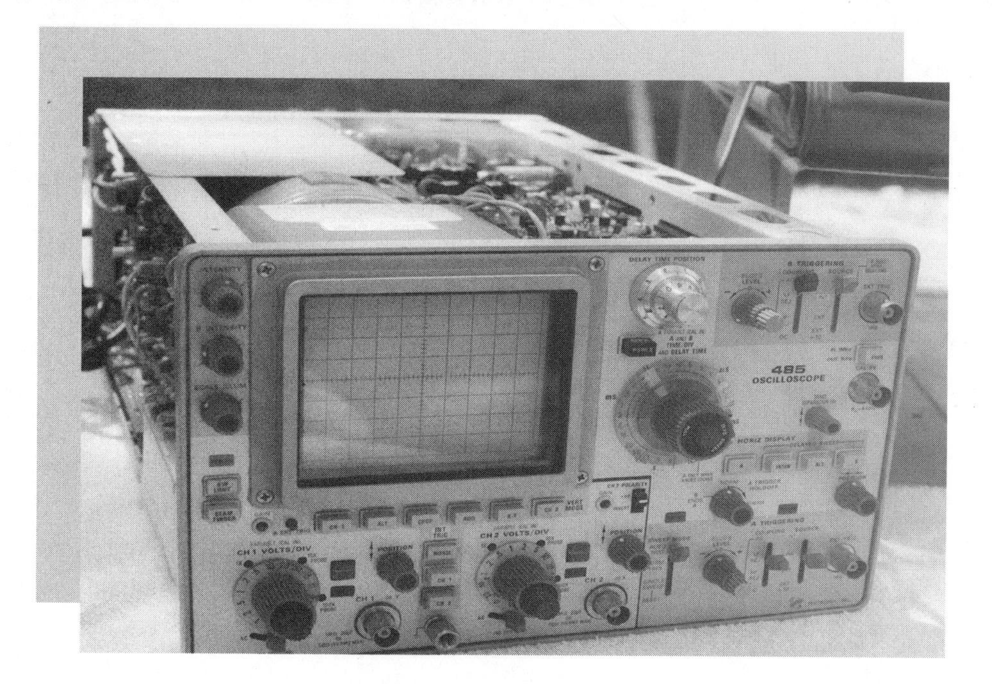

APPENDIX C

QUIZ ANSWERS

QUIZ 1:

1: A
2: B
3: A
4: D
5: A
6: C
7: A
8: B
9: C
10: B
11: B
12: B
13: A
14: C
15: C

QUIZ 2:

1: A
2: B
3: A
4: C
5: C
6: C
7: C
8: A
9: A
10: C
11: B
12: A
13: B
14: A
15: D

QUIZ 3:

1: C	8: A	15: D
2: B	9: B	
3: A	10: A	
4: C	11: C	
5: A	12: C	
6: A	13: A	
7: A	14: A	

QUIZ 4:

1: C	8: B	15: B
2: A	9: B	
3: A	10: B	
4: A	11: A	
5: B	12: A	
6: A	13: A	
7: C	14: B	

QUIZ 5:

1: B	8: B	15: B
2: A	9: B	
3: B	10: C	
4: A	11: A	
5: B	12: B	
6: A	13: C	
7: C	14: B	

QUIZ 6:

1: B	8: B	15: B
2: B	9: C	
3: A	10: B	
4: A	11: A	
5: A	12: A	
6: B	13: A	
7: A	14: B	

QUIZ 7:

1: A	8: A	15: B
2: B	9: B	
3: B	10: A	
4: A	11: B	
5: A	12: A	
6: D	13: A	
7: C	14: B	

QUIZ 8:

1: A	8: B	15: A
2: A	9: A	
3: A	10: C	
4: A	11: B	
5: B	12: C	
6: B	13: C	
7: A	14: B	

QUIZ 9:

1: B
2: B
3: B
4: B
5: B
6: A
7: A
8: C
9: A
10: B
11: A
12: A
13: B
14: A
15: C

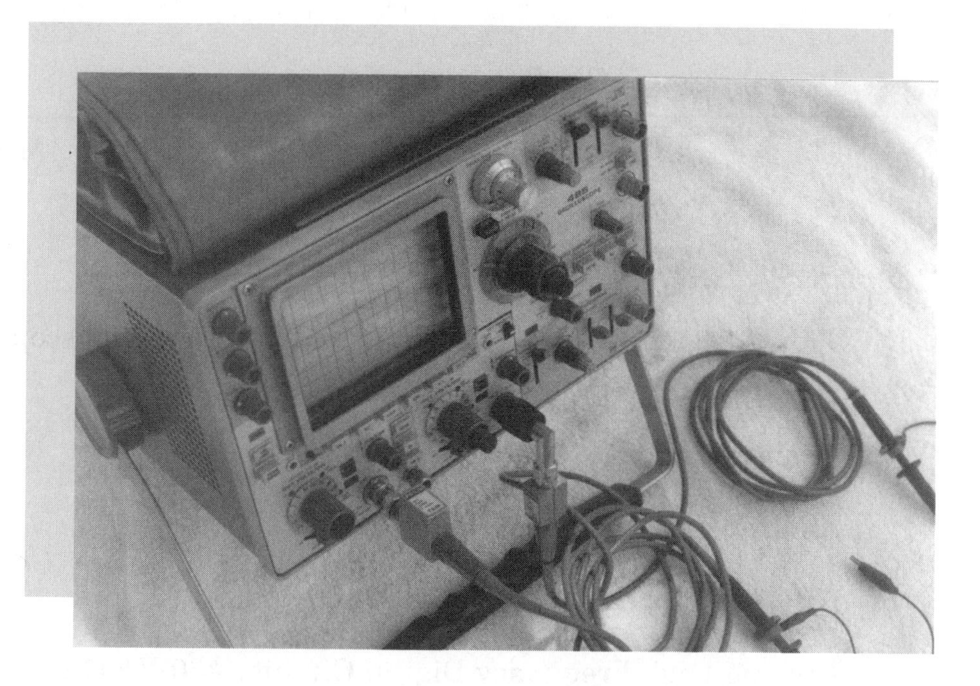

APPENDIX D

REFERENCES:

The Electrical Engineering Handbook. CRC PRESS.

Reference Data for Engineers, Seventh Edition. Howard W. Sams & Co.

Oscilloscopes: How to Use Them, How They Work, Fourth Edition. Newnes.

Digital Storage Oscilloscopes. Newness.

The Complete Book of Oscilloscopes, Second Edition. McGraw Hill.

Hands-On Guide to Oscilloscopes. McGraw Hill.

Troubleshooting With Your Triggered-Sweep Oscilloscope. McGraw Hill

Basic Electronics. Master Publishing, Inc.

Additional Articles & References

ABC's of Probes #60W-6053-5. Tektronix.

Probing High Frequency Digital Circuity #60W-8412-0. Tektronix.

The Effects of Probe Input Capacitance on Measurement Accuracy #60W8910. Tektronix.

The XYZ's of Oscilloscopes. Tektronix.

Differential Oscilloscope Measurements #51W-10540-0. Tektronix.

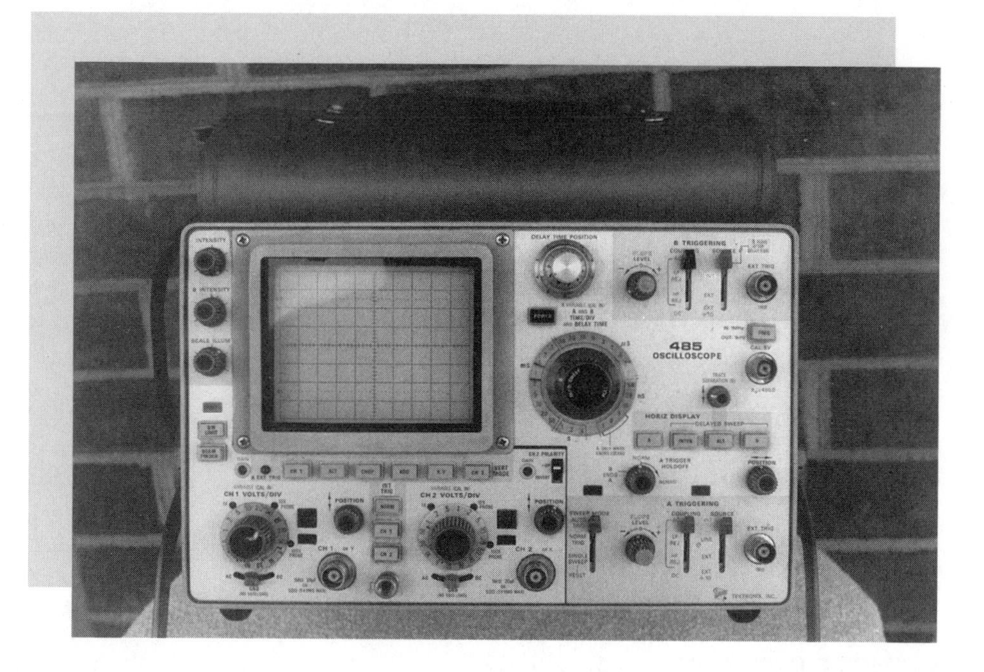
DELAY TIME POSITION
B TRIGGERING
485
OSCILLOSCOPE
HORIZ DISPLAY
CH 1 VOLTS/DIV
CH 2 VOLTS/DIV
A TRIGGERING

INDEX

A

B

C

D

E

F

G

H

M

N

O

P

T

U

V

W

X

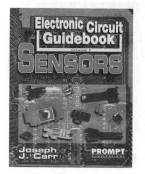

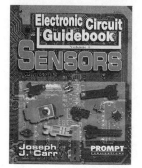

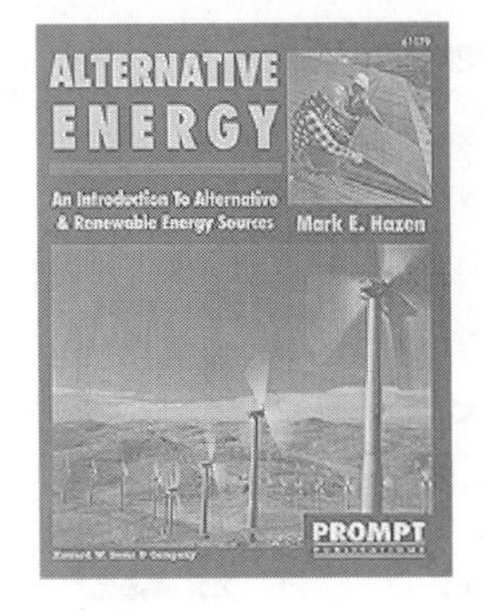

Alternative Energy

Mark E. Hazen

This book is designed to introduce readers to the many different forms of energy mankind has learned to put to use. Generally, energy sources are harnessed for the purpose of producing electricity. This process relies on transducers to transform energy from one form into another. *Alternative Energy* will not only address transducers and the five most common sources of energy that can be converted to electricity, it will also explore solar energy, the harnessing of the wind for energy, geothermal energy, and nuclear energy.

This book is designed to be an introduction to energy and alternate sources of electricity. Each of the nine chapters are followed by questions to test comprehension, making it ideal for students and teachers alike. In addition, listings of World Wide Web sites are included so that readers can learn more about alternative energy and the organizations devoted to it.

Professional Reference
320 pages • Paperback • 7-3/8 x 9-1/4"
ISBN: 0-7906-1079-5 • Sams: 61079
$18.95 ($25.95 Canada) • October 1996

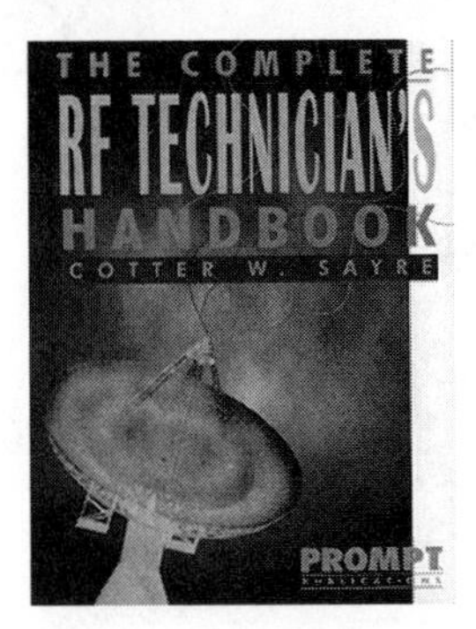

The Complete RF Technician's Handbook

Cotter W. Sayre

The *Complete RF Technician's Handbook* will furnish the working technician or student with a solid grounding in the latest methods and circuits employed in today's RF communications gear. It will also give readers the ability to test and troubleshoot transmitters, transceivers, and receivers with absolute confidence. Some of the topics covered include reactance, phase angle, logarithms, diodes, passive filters, amplifiers, and distortion. Various multiplexing methods and data, satellite, spread spectrum, cellular, and microwave communication technologies are discussed.

Cotter W. Sayre is an electronics design engineer with Goldstar Development, Inc., in Lake Elsinore, California. He is a graduate of Los Angeles Pierce College and is certified by the National Association of Radio and Telecommunications Engineers, as well as the International Society of Electronics Technicians.

Professional Reference
281 pages • Paperback • 8-1/2 x 11"
ISBN: 0-7906-1085-X • Sams: 61085
$24.95 ($33.95 Canada) • July 1996

CALL 1-800-428-7267 TODAY FOR THE NAME OF YOUR NEAREST PROMPT PUBLICATIONS DISTRIBUTOR

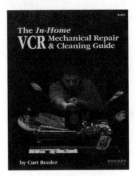

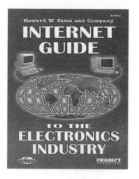